Sven Bodo Wirsing

Maximal nilpotent subalgebras I

Nilradicals and Cartan subalgebras
in associative algebras

With 428 exercises

Anchor Academic
Publishing

Wirsing, Sven Bodo: Maximal nilpotent subalgebras I: Nilradicals and Cartan subalgebras in associative algebras. With 428 exercises, Hamburg, Anchor Academic Publishing 2016

Buch-ISBN: 978-3-96067-103-9
PDF-eBook-ISBN: 978-3-96067-603-4
Druck/Herstellung: Anchor Academic Publishing, Hamburg, 2016
Covermotiv: designed by freepik.com

Bibliografische Information der Deutschen Nationalbibliothek:
Die Deutsche Nationalbibliothek verzeichnet diese Publikation in der Deutschen Nationalbibliografie; detaillierte bibliografische Daten sind im Internet über http://dnb.d-nb.de abrufbar.

Bibliographical Information of the German National Library:
The German National Library lists this publication in the German National Bibliography. Detailed bibliographic data can be found at: http://dnb.d-nb.de

All rights reserved. This publication may not be reproduced, stored in a retrieval system or transmitted, in any form or by any means, electronic, mechanical, photocopying, recording or otherwise, without the prior permission of the publishers.

Das Werk einschließlich aller seiner Teile ist urheberrechtlich geschützt. Jede Verwertung außerhalb der Grenzen des Urheberrechtsgesetzes ist ohne Zustimmung des Verlages unzulässig und strafbar. Dies gilt insbesondere für Vervielfältigungen, Übersetzungen, Mikroverfilmungen und die Einspeicherung und Bearbeitung in elektronischen Systemen.

Die Wiedergabe von Gebrauchsnamen, Handelsnamen, Warenbezeichnungen usw. in diesem Werk berechtigt auch ohne besondere Kennzeichnung nicht zu der Annahme, dass solche Namen im Sinne der Warenzeichen- und Markenschutz-Gesetzgebung als frei zu betrachten wären und daher von jedermann benutzt werden dürften.

Die Informationen in diesem Werk wurden mit Sorgfalt erarbeitet. Dennoch können Fehler nicht vollständig ausgeschlossen werden und die Diplomica Verlag GmbH, die Autoren oder Übersetzer übernehmen keine juristische Verantwortung oder irgendeine Haftung für evtl. verbliebene fehlerhafte Angaben und deren Folgen.

Alle Rechte vorbehalten

© Anchor Academic Publishing, Imprint der Diplomica Verlag GmbH
Hermannstal 119k, 22119 Hamburg
http://www.diplomica-verlag.de, Hamburg 2016
Printed in Germany

For my beloved mother

A mother is special, she's more than a friend.
Whenever you need her, she'll give you a hand.
She'll lead you and guide you in all that you do.
Try all that she can just to see you get through.
Good times and bad times, she's there for it all.
Say head up, be proud, and always stand tall.
She'll love you through quarrels and even big fights,
or heart to heart chats on cold lonely nights.
My mother's the greatest that I've ever known,
I think God made my mother like He'd make his own.
A praiser, a helper, an encourager too,
nothing in this world that she wouldn't do.
To help us succeed she does all that she can,
raised a young boy now into a man.
I want to say thank you for all that you do,
please always know mom, that I love you.

(A true angel by George W. Zellars, February 2006)

Contents

	Introduction	**7**
1	**Natural examples**	**13**
2	**Finite subgroups of fields and division algebras**	**17**
	2.1 Finite subgroups of fields	17
	2.2 Results of Wedderburn, Amitsur and Herstein about division algebras	18
	2.3 Exercises	22
3	**Normal and subnormal subgroups in unit groups of division algebras**	**25**
	3.1 The theorem of Cartan-Brauer-Hua	25
	3.2 The theorem of Scott	27
	3.3 The theorem of Stuth	28
	3.4 Exercises	31
4	**Nilradicals of Lie algebras associated to associative algebras**	**33**
	4.1 Solvable algebras	34
	4.1.1 Generalized Jordan decomposition	34
	4.1.2 The nilradical	35
	4.2 Standard examples of solvable associative algebras	37
	4.2.1 Solvable group algebras	37
	4.2.2 Triangular matrices	41
	4.2.3 Solomon algebras in characteristic zero	42
	4.2.4 Solomon-Tits algebras	42
	4.3 Herstein's analysis for simple associative rings	45
	4.4 Simple and semisimple algebras	46
	4.4.1 Simple algebras	46
	4.4.2 Direct products and semisimple algebras	48
	4.5 Associative algebras with separable factor algebra by its nilradical	51
	4.6 Compatibility of the Lie nilradical with special constructions of algebras	53

	4.6.1	Subalgebras	53
	4.6.2	Right and left ideals	53
	4.6.3	Ideals	54
	4.6.4	Factor algebras	54
	4.6.5	The opposite algebra	54
	4.6.6	Matrix algebras	54
	4.6.7	Adjunction of a unit	55
	4.6.8	Tensor products	56
4.7	The group algebra once again		56
4.8	Open-ended questions and exercises		59

5 Cartan subalgebras in Lie algebras associated to associative algebras   65

5.1	Cartan subalgebras are associative solvable subalgebras . . .		65
	5.1.1	Associative structure of Cartan subalgebras	66
	5.1.2	Open-ended questions and exercises	68
5.2	Maximal tori and Cartan subalgebras		71
	5.2.1	Maximal tori	71
	5.2.2	Maximal tori of group algebras based on dihedral and quaternion groups	72
	5.2.3	Cartan subalgebras	73
	5.2.4	Cartan subalgebras of group algebras based on dihedral and quaternion groups	75
	5.2.5	Maximal tori and Cartan subalgebras of Solomon algebras, Solomon-Tits algebras and triangular matrices	76
	5.2.6	Open-ended questions and exercises	79
5.3	Division algebras		80
	5.3.1	Central division algebras	81
	5.3.2	Non-central division algebras	82
	5.3.3	Open-ended questions and exercises	85
5.4	Quaternion algebras		86
	5.4.1	The case characteristic $\neq 2$	86
	5.4.2	Examples for the associative conjugacy	86
	5.4.3	Remark on the Lie conjugacy	87
	5.4.4	The case characteristic 2	87
	5.4.5	Open-ended questions and exercises	90
5.5	Solvable algebras		90
	5.5.1	Unitary solvable algebras	91
	5.5.2	The quasi regular group and non-unitary solvable associative algebras	93
	5.5.3	Solvable group algebras	96
	5.5.4	Solvable group algebras of dihedral groups	97
	5.5.5	Solvable group algebras of quaternion groups	100
	5.5.6	Splitting solvable algebras	101

	5.5.7	Open-ended questions and exercises 105
5.6	Lie nilpotent associative algebras 107	
	5.6.1	Lie nilpotency 107
	5.6.2	Nilpotent group of units 110
	5.6.3	Group algebras and Lie nilpotency 112
	5.6.4	Exercises 114
5.7	Simple, semisimple and separable algebras 116	
	5.7.1	Central-simple algebras 116
	5.7.2	Simple algebras 119
	5.7.3	Semisimple and separable algebras 119
	5.7.4	Open-ended questions and exercises 122
5.8	Basic algebras 125	
	5.8.1	Characterizations of reduced algebras 125
	5.8.2	Maximal solvable subalgebras 126
	5.8.3	Cartan subalgebras 127
	5.8.4	Maximal nilpotent subalgebras 128
	5.8.5	Reduced group algebras: the semisimple case 129
	5.8.6	Reduced group algebras: the modular case 133
	5.8.7	An example of Benjamin Steinberg 134
	5.8.8	Open-ended questions and exercises 137
5.9	Associative algebras with separable factor algebra by its nil-radical 143	
	5.9.1	A description by radical complements 143
	5.9.2	A strategy for the determination of Cartan subalgebras 144
	5.9.3	Solvable algebras revised 144
	5.9.4	Group algebras for dihedral groups 144
	5.9.5	Open-ended questions and exercises 147
5.10	Natural determination of Cartan subalgebras 148	
	5.10.1	Sub- and factor structures 148
	5.10.2	Direct products 150
	5.10.3	Adjunction of a unit 150
	5.10.4	Cyclic algebras 151
	5.10.5	Tensor products and extension of the base field 151
	5.10.6	Matrix algebras 151
	5.10.7	Group algebras 152
	5.10.8	Open-ended questions and exercises 154

6 Summation formulas for the dimension of maximal tori in group algebras   157
6.1 The semisimple case 157
6.2 Upper and lower bounds 158
6.3 Special classes of groups 160
6.4 The modular case 174
6.5 Open-ended questions and exercises 178

7	**Invariants**	**189**
	7.1 Dimension formula for maximal tori	189
	7.2 Cartan subalgebras	191
	7.3 Open-ended questions and exercises	206
8	**Outlook on series II**	**213**
A	**Derived algebras**	**215**
	A.1 Definition and initial properties	215
	A.2 Structural properties	217
	A.3 Lie nilpotency	217
	A.4 Isomorphism of derived algebras	218
	A.5 Open-ended questions and exercises	223

List of figures   **227**

Bibliography   **227**

Index   **234**

Introduction

Maximal nilpotent are Cartan-subalgebras
as well as the nilradical.
Both will be studied in the magnificient sal
of associated Lie algebras.

(Sven Wirsing, December 2015)

Within the theory of Lie algebras Cartan subalgebras play an important role for the classification of semisimple Lie algebras as well as within the theory of symmetric spaces.

During my time of studying at the Christian-Albrechts-Universität of Kiel Salvatore Siciliano presented his researches in the Oberseminar Algebrentheorie about Cartan subalgebras in Lie algebras associated to associative algebras. His presentation was the starting point for me to study maximal nilpotent substructures in associated Lie algebras of associative algebras. In this work we will present his theory of Cartan subalgebras and enhance it to some special associative algebras (e.g. basic algebras, division algebras, algebras with separable factor algebra by its nilradical). In addition, a second maximal nilpotent substructure is analyzed, its the so-called nilradical of a Lie algebra.

The first chapter introduces some special associative and Lie algebras, monoids and groups. They will be important to visualize and illustrate the general theorems proven within this work. Some applications are also transferred to the exercises at the end of each section or chapter. There are some exercises included enhancing the theory presented so far such that the reader gets a deeper insight. In addition, at the beginning of each exercise series some open-ended topics are included which can be used by the reader – and also by the author – to do additional researches within this theory. The author has included some (manually created) graphics – mostly so called Hasse diagrams – to visualize the results of each section or chapter.

Within chapter 2 basic results about finite subgroups of fields and division algebras are summarized. Some will be proven in details, others will be just presented without a proof. They will play a role later on in the next chapters of this work such that their understanding leads to a better insight of the latter results. In addition, the author includes some proofs of these basic results because of personal interest on the proofs itself. The summary will include the proof that finite subgroups of fields are cyclic, the theorem of Wedderburn about finite division algebras as well as results of Herstein and Amitsur about the classification of finite subgroups of division algebras.

Likewise structured is chapter 3. This chapter focusses on the normal and subnormal subgroup structure of division algebras. We will prove the theorem of Cartan-Brauer-Hua about normal subgroups of division algebras and the theorem of Scott about solvable group of units of division algebras. Finally, the theorem of Stuth about subnormal subgroups is presented (without proving it) enhancing the theorem of Scott.

For an associative algebra the associated Lie algebra can be derived in a natural way. In chapter 4 we analyze the nilradical – the greatest nilpotent ideal – of that Lie algebra and focus our analysis on its associative structure. For this, the center and the nilradical of the associative algebra are of importance: the nilradical is the sum of these two associative substructures. In particular, its an associative subalgebra. For this theorem we assume that the factor algebra by the nilradical of the associative algebra is separable and thus we can use the theorem of Wedderburn-Malcev within the proof.
The analysis begins by determining the nilradical in the case of a solvable associative algebra. For this, results of the so-called generalized Jordan decomposition are used. We demonstrate the theorem on special solvable group algebras based on dihedral and quaternion groups, on the Solomon algebras in characteristic zero, on the Solomon-Tits algebras and on the algebras of upper and lower triangular matrices over an arbitrary field.
In a second step some results of Herstein about simple rings and their associated Lie ring are transferred to simple and semisimple algebras: we will prove that the nilradical is identical to the greatest solvable Lie ideal – the so-called solvable radical. Both structures are identical to the center of the associative algebra. For the semisimple case it is proven that the Lie nilradical of direct products is the direct product of the Lie nilradicals of the corresponding components: there are no diagonals possible.
Both results – for solvable and semisimple associative algebras – are used to determine the nilradical for arbitrary associative algebras.
The chapter is finalized to apply and enhance the theorem for algebra constructions like the tensor product, the adjunction of a unit and matrix algebras over algebras. The idea is to determine the Lie nilradical by the components of the algebra constructions, like by the factors for the tensor product. We will give proofs or counterexamples for these constructions

with respect to this question.

In the previous chapter we have deeply analyzed the Lie nilradical of an associative algebra with respect to its associative structure. The Lie nilradical is a maximal nilpotent substructure, and the Cartan subalgebras are maximal nilpotent, too. They are in focus of the next chapter. They are defined as being nilpotent and self-normalizing Lie subalgebras. The aim of this chapter is the same as for Lie nilradical: their determination and the description of their associative structure. Some results of this chapter are based on an article of Salvatore Siciliano [59], others are enhancements of his theory to other classes of associative algebras like division algebras, simple, semisimple and separable associative algebras, reduced associative algebras or associative algebras with separable factor algebra by their nilradical. Standard examples are investigated in details, in particular group algebras, lower and upper triangular matrices and Solomon-(Tits) algebras for illustrating the developed theory.

The main result of this chapter is the 1:1-connection between maximal tori (maximal commutative separable subalgebras) und Cartan subalgebras. Centralizing maximal tori is a bijection between these structures. The inverse calculates for every Cartan subalgebras a maximal torus by creating the set of fully separable elements of the Cartan subalgebra.

In some cases both sets – maximal tori and Cartan subalgebras – are identical, like for separable associative algebras. Central divisions algebras are separable, too, and we prove a theorem of Salvatore Siciliano (in a different way) that maximal tori and Cartan subalgebras are exactly the maximal separable subfields. We enhance the theorem by proving that these are exactly the separable maximal subfields which is also an alternative proof of a theorem of Emmy Noether. In particular, it is proven that all maximal tori = Cartan subalgebras have the same dimension and are isomorphic as Lie algebras. This theorem is transferred to non-central division algebras.

Solvable associative algebras have the property that maximal tori are exactly the radical complements if the factor algebra by its associative nilradical is separable. This result – proven by Thorsten Bauer in his dissertation [4] and by Salvatore Siciliano in [59] – is proven by a different approach and revised later on in the second to last section of this chapter. As a consequence of our main theorem about Cartan subalgebras and the theorem of Wedderburn-Malcev all maximal tori and Cartan subalgebras are conjugated, and the Cartan-subalgebras are exactly the centralizers of the radical complements. For basic algebras we transfer the determination of Cartan subalgebras to Cartan subalgebras of maximal solvable substructures. These maximal ones are describable as direct sums of maximal tori and the associative nilradical. The centralizers of the maximal tori of the underlying algebra are identical to the centralizer within these maximal solvable subalgebras. Afterwards we focus on reduced group algebras. In the modular case the terms basic and solvable are equivalent. For semisimple group algebras the situation is more complex: the group is hamiltonian and the equation $a^2 + b^2 + 1 = 0$ has no solution in special field extension based on roots of unities. Finally, we determine the dimension of the Cartan subalgebras for these group algebras based on the results of chapter 6.

In the second to last section we analyze how the determination of Cartan subalgebras can be done based on separable radical complements. The maximal tori of the radical complement and of the whole algebra are identically. For separable radical complements maximal tori and Cartan subalgebras are identically, too. The centralizers of them are exactly the Cartan subalgebra of the underlying algebra. Based on this result a strategy is developed for determining Cartan subalgebras. For solvable algebras this strategy is used and the determination of Cartan subalgebras is revised in a more transparent way. We apply this strategy also on group algebras of dihedral groups. The chapter is finalized to apply and enhance the theorem for Cartan subalgebras for algebra constructions like the tensor product, the adjunction of a unit and matrix algebras over algebras. The idea is to determine the Lie nilradical by the components of the algebra constructions, like by the factors for the tensor product. We will give proofs or counterexamples for these constructions with respect to this question.

The next chapter is dedicated to the dimension of maximal tori in group algebras. We begin this chapter by proving a result of Salvatore Siciliano connecting this dimension to the sum of degrees of all irreducible complex characters for semisimple group algebras. This sum is identical for all fields such that the group algebra is semisimple. We use this result and some classical and modern results about that sum within the character theory of finite groups to bound this dimension – like by the number of involutions, by the order of the radical, by the order of abelian subgroups and by the maximal degree – and determine this sum for several classes of groups – like for Frobenius groups, for direct products, for extra special p-groups, for diverse linear groups, for ambivalent groups such as dihedral and symmetric groups, for meta-cyclic groups, for p-groups, for nilpotent groups and for minimal non-abelian p-groups.

Within chapter 7 we focus on the question whether the dimensions of the maximal tori and of the Cartan subalgebras are unique for associated Lie algebras of finite-dimensional associative unital algebras. For maximal tori we give a positive answer to this question for associative algebras with separable factor algebra by its nilradical by calculating this dimension explicitly. The answer for the Cartan subalgebras is positive, too. In characteristic zero we derive this result by using a classical result on Cartan subalgebras over algebraically closed fields. In the modular case we begin the analysis by proving the uniqueness for associated Lie algebras based on solvable finite-dimensional associative algebras, for separable associative algebras and for finite-dimensional associative algebras possessing a central nilradical. The general case is derived by using a result of Premet (which was later proven by Farnsteiner) for restricted Lie algebras over algebraically closed fields in positive characteristic and by using the result on the dimension for maximal tori. In general, the dimension of Cartan subalgebras can differ for restricted Lie algebras. By using a second approach we extend our theorem for the uniqueness of the dimension of Cartan subalgebras to the solvable and nilpotency class. For this, we prove that all maximal tori and Cartan subalgebras of Lie algebras associated to finite-dimensional associative algebras over an arbitrary algebraically closed field are conjugated. We demonstrates these three invariants – dimension, nilpotency and solvable class – by calculating them for group algebras based on dihedral and quaternion groups.

Chapter 8 is an outlook on the second series about maximal nilpotent substructures. We will focus on the solvable case of an associative algebra in more details as in this first volume. For this, we will extend the topic to all maximal nilpotent substructures and to the connection to the maximal nilpotent subgroups of their group of unit. A graphic illustrates the problems analyzed in series II.

Within the appendix we classify a special class of algebras and analyze their Lie nilpotency. This class of algebras was in focus of the diploma thesis of Armin Jöllenbeck.

Chapter 1

Natural examples

This chapter has a preliminary function by summarizing those monoids, groups, associative and Lie algebras which will arise in this work. They will be used for examples of the proven theorems as well as for exercises in which the reader shall apply the results.

Groups and monoids

Let $n \in \mathbb{N}$, N be a set, M a monoid, G a group, A an associative unitary algebra and q a prime power. We will focus on the following groups and monoids:

- $\mathbb{N}$ - natural numbers
- $\mathbb{N}_0$ - natural numbers containing zero
- $(P(N); \cap)$ - power set of N with operation $\cap$
- $(P(N); \cup)$ - power set of N with operation $\cup$
- $(P(N); \delta)$ - power set of N with operation δ - symmetric difference
- $(P(M); \cdot)$ - power set of M with complex product $\cdot$ as operation
- $(P(G); \cdot)$ - power set of G with complex product $\cdot$ as operation
- D_{2n} - dihedral group of order $2n$
- Q_{4n} - quaternion group of order $4n$
- SD_{2^n} - semi-dihedral group of order 2^n
- S_n - symmetric group of degree n
- A_n - alternating group of degree n

- $GL(n,q)$ - general linear group of degree n over $GF(q)$
- $SL(n,q)$ - special general linear group of degree n over $GF(q)$
- $PSL(n,q)$ - projective special general linear group of degree n over $GF(q)$
- $SP(2n,q)$ - symplectic group of degree $2n$ over $GF(q)$
- $GSP(2n,q)$ - general similitudes group
- $U(n,q)$ - unitary group of degree n over $GF(q)$
- C_n or Z_n - cyclic group of order n
- $E(A)$ - group of units of A
- $Q(A)$ - quasiregular group of A
- $\times$ - direct products of groups
- $\wr$ - regular wreath product of groups
- $\ltimes$ - semidirect product of groups.

General constructions of algebras

Let A be an algebra, K a field, G a group, I an ideal, M a monoid, $n \in \mathbb{N}$ and $T \subseteq A$. The following general constructions of algebras will be used:

- $\otimes$ - tensor product of algebras
- $\times$ - direct products of algebras
- $\oplus$ - direct sum of algebras
- A/I - factor algebra of A by the ideal I
- KG - group algebra of the group G and the field K
- KM - monoid algebra of the monoid M and the field K
- $A^{n \times n}$ - algebra of $n \times n$-matrices over A
- A° - associated Lie algebra of A
- $\langle T \rangle_K$ - K-linear span of T
- $\langle T \rangle_\mathcal{A}$ - subalgebra generated by T
- $\langle T \rangle_{\mathcal{A}_1}$ - unital subalgebra generated by T
- A^K - adjunction of a unit to A

- A^{op} or A^- - inverse or opposite algebra of A
- $(A \times A; \odot)$ - zero extension of A
- $gl(n, K)$ - identical to $(K^{n \times n})^\circ$
- eAe - identical to $\{eae \mid a \in A\}$ for an idempotent e
- $Aug(KG)$ - augmentation ideal of KG.

Commutative algebras

The following commutative algebras will appear:
- $\mathbb{Z}$ - the set of integers
- $K[t]$ - polynomial algebra over K in one variable t.

Fields and skew fields

Let p be a prime number, $n \in \mathbb{N}$ and $(K; L)$ a field extension. We will focus on the following fields, skew fields and elements:
- $\mathbb{Q}$ - rational number field
- $\mathbb{R}$ - real number field
- $\mathbb{C}$ - complex number field
- $\mathbb{H}$ - real quaternion algebra
- $GF(p^n)$ - finite field with p^n elements
- $GF(q)$ - notation for $GF(p^n)$ and $q = p^n$
- $A(a, b)$ - generalized quaternion algebra
- $K(a)$ - smallest subfield in L containing a and K
- ω_d - primitive dth root of unity
- cyclic division algebras.

(Central) - simple associative algebras

Let K be a field, D a division algebra and $n \in \mathbb{N}$. We will use the following (central)-simple associative algebras:
- $K^{n \times n}$ - $n \times n$-matrices over K
- $D^{n \times n}$ - $n \times n$-matrices over D
- $A(a, b)$ - generalized quaternion algebra.

Semisimple associative algebras

We will use the following semisimple associative algebras:

- $\times$ - direct products of simple algebras
- $A/rad(A)$ - the factor algebra by the nilradical of an associative algebra.

Nilpotent associative algebras

Let A be an associative algebra, K a field, p a prime number, $n \in \mathbb{N}$ and G a p-group. We will focus on the following nilpotent associative algebras:

- $rad(A)$ - nilradical of A
- $J(A)$ - Jacobson radical of A
- $s\delta_{u,n}$ - algebra of strict lower triangular matrices of $K^{n \times n}$
- $s\delta_{o,n}$ - algebra of strict upper triangular matrices of $K^{n \times n}$
- $Aug(KG)$ - augmentation ideal of KG based on a p-group G and $char(K) = p$.

Solvable associative algebras

Let $n \in \mathbb{N}$, p a prime number, G a finite group and K be a field. We will focus on the following solvable associative algebras:

- $K\Pi_n$ - Solomon-Tits algebra (see e.g. [76])
- D_n - Solomon algebra in the case $char(K) = 0$ (see e.g. [4])
- $\delta_{u,n}$ - algebra of lower triangular matrices of $K^{n \times n}$
- $\delta_{o,n}$ - algebra of upper triangular matrices of $K^{n \times n}$
- KG - group algebra based on: $char(K) = p$ and G possesses a normal p-Sylow subgroup with an abelian p'-Hall subgroup.

Chapter 2

Finite subgroups of fields and division algebras

In this chapter we summarize some results of finite subgroups in unit groups of fields and division algebras. For some of them we provide a proof, for the others we reference the corresponding literature. We will use some of these results in the next chapters. Therefor these results provide the reader a deeper insight for understanding these results. In addition, this chapter is included on personal interest of the author for the proofs of these results.

2.1 Finite subgroups of fields

By $E(A)$ and $K[t]$ we denote the group of units of an associative algebra A and the algebra of polynomials over a ring K based on the single variable t. For a group G and an element g of G let $o(g)$ (more exact: $o_G(g)$) the order of g in G.

The following theorem is proven by various arguments within the literature. It is unknown which mathematician provided the first proof of this result. Our variant is based on the main theorem on finite abelian groups.

Theorem 1 *Every finite subgroup of the group of units of a field is cyclic. In particular, the group of units of a finite field is cyclic.*

Proof. Let K be a field and U a finite subgroup of $E(K)$. By using the main theorem on finite abelian groups we decompose U in cyclic groups of prime power order:

$$U = (G_{1,1} \times \cdots \times G_{1,s_1}) \times \cdots \times (G_{r,1} \times \cdots \times G_{r,s_r}).$$

In this decomposition all groups $G_{i,j}$ are of prime power order with respect to the prime number p_i. We arrange the product such that $G_{i,1}$ is the greatest factor within $G_{i,1} \times \cdots \times G_{i,r_i}$. For every i let g_i a generator of $G_{i,1}$.

We focus on the element $g := g_1 \cdots g_r$. g is of order $o(g) = o(g_1) \cdots o(g_r)$ because all prime numbers $p_1, \cdots, p_r$ are distinct. For every $u \in U$ the identity $u^{o(g)} = 1$ is valid.

All elements of U are roots of the polynomial $t^{o(g)} - 1$, and there are at most $o(g)$ distinct roots. Hence we derive $\mid U \mid \leq o(g)$. All $o(g)$-powers of g are distinct. Therefor U is exactly the set of these powers of g. We conclude that U is cyclic and generated by g. ⋄

2.2 Results of Wedderburn, Amitsur and Herstein about division algebras

An unitary algebra is an algebra with a unit. An unital subalgebra of an unitary algebra is a subalgebra containing the unit element of the global algebra. Hence a unital subalgebra is unitary. An unitary subalgebra is a subalgebra which is unitary as an algebra. An unitary subalgebra does not need to be unital as its unitary unit could differ from the unit element of the global algebra. Its unit element is an idempotent of the global algebra. The center of A is denoted by $Z(A)$.

Let G be a group, T a subset of G and $g \in G$. By ^{g} we symbolize the conjugation with g and by $C_G(T)$ resp. $N_G(T)$ the centralizer resp. normalizer of T in G.

Our next focus is the proof of a theorem of Wedderburn about finite division algebras. For this proof we need the following two propositions.

Proposition 1 *Let D be a K-division algebra and T be a unital finite-dimensional subalgebra of D. Then T is a division algebra, too.*

Proof. Let $t \in T$ and assume $t \neq 0$. We consider the right and the left multiplication with t on T. Both functions are injective because D is a division algebra. Hence - using the finite dimension of T - they are surjective, too. In particular, 1 has a pre-image with respect to these functions. Both pre-images are the inverse of t and therefor contained in T. ⋄

Proposition 2 *Let G be a finite group and U be a subgroup of G. The the U and G are equal if and only if G is the union of all G-conjugate subgroups of U.*

Proof. If U is a normal subgroup the statement is true. Let U be a non-normal subgroup of G. Hence the statement $G > N_G(U)$ is true. The number of conjugates of U is exactly the index of the normalizer of U in G which is $\frac{|G|}{|N_G(U)|}$. All conjugated of U have at least the unit element in common. Therefor we conclude:

$$\mid \bigcup_{g \in G} U^g \mid \leq 1 + \frac{|G|}{|N_G(U)|} \cdot (\mid U \mid -1).$$

The right hand side is – because of $U \leq N_G(U)$ – not greater than

$$1+ \mid G \mid - \frac{|G|}{|N_G(U)|}.$$

By using $G > N_G(U)$ we derive that this value is smaller than $\mid G \mid$. ◇

We will prove the following theorem by usage of the theory of central-simple associative division algebras. For this, let $ind(D)$ (more exact: $ind_K(D)$) the index of a central-simple finite-dimensional associative unitary K-division algebra which is the unique dimension of all maximal subfields of D. A good introduction to this theory can be found [49] and in [39].

Theorem 2 *(Wedderburn) Every finite division algebra is a field. In particular, its group of units is cyclic.*

Proof. Let D be a finite division algebra and $K := Z(D)$. K is a field and D a central-simple finite-dimensional associative unitary K-division algebra. All maximal subfields have the same dimension $ind_K(D)$. Hence – by using the finiteness of D – they are of the same order. Based on the finite field theory we know that all maximal subfields are isomorphic. Now we use the theorem of Skolem-Noether[1] and conclude that all maximal subfields are conjugated. Every element d of D is contained in a maximal subfield of D

---

[1]Thoralf Albert Skolem (born 23 May 1887, died 23 March 1963) was a Norwegian mathematician who worked in mathematical logic and set theory. Although Skolem's father was a primary school teacher, most of his extended family were farmers. Skolem attended secondary school in Kristiania (later renamed Oslo), passing the university entrance examinations in 1905. He then entered Det Kongelige Frederiks Universitet to study mathematics, also taking courses in physics, chemistry, zoology and botany. In 1909, he began working as an assistant to the physicist Kristian Birkeland, known for bombarding magnetized spheres with electrons and obtaining aurora-like effects; thus Skolem's first publications were physics papers written jointly with Birkeland. In 1913, Skolem passed the state examinations with distinction, and completed a dissertation titled Investigations on the Algebra of Logic. He also traveled with Birkeland to the Sudan to observe the zodiacal light. He spent the winter semester of 1915 at the University of Göttingen, at the time the leading research center in mathematical logic, metamathematics, and abstract algebra, fields in which Skolem eventually excelled. In 1916 he was appointed a research fellow at Det Kongelige Frederiks Universitet. In 1918, he became a Docent in Mathematics and was elected to the Norwegian Academy of Science and Letters. Skolem did not at first formally enroll as a Ph.D. candidate, believing that the Ph.D. was unnecessary in Norway. He later changed his mind and submitted a thesis in 1926, titled Some theorems about integral solutions to certain algebraic equations and inequalities. His notional thesis advisor was Axel Thue, even though Thue had died in 1922. In 1927, he married Edith Wilhelmine Hasvold. Skolem continued to teach at Det kongelige Frederiks Universitet (renamed the University of Oslo in 1939) until 1930 when he became a Research Associate in Chr. Michelsen Institute in Bergen. This senior post allowed Skolem to conduct research free of administrative and teaching duties. However, the position also required that he reside in Bergen, a city which then lacked a university and hence had no

because the subalgebra of D generated by $\{d, 1\}$ is a subfield of D (see proposition 1). Therefor D is the union of all maximal subfields of D. From this we derive that $E(D)$ is the union of all groups of units of all maximal subfields and that these subgroups are conjugated. We can apply proposition 2 and conclude that D and one maximal subfield of D are identical. The proof is complete and the add-on is a consequence of this result and of theorem 1.⋄

Let V be a K-linear space and T a subset of V. By $\langle T \rangle_V$ we denote the K-linear span of T in V. $GF(p^n)$ resp. $GF(q)$ symbolize a finite field of order p^n resp. q (Galois field).

By usage of our previous results we derive two theorems proven by Herstein:

Theorem 3 *(Herstein) Every finite abelian subgroup of a division algebra is cyclic.*

Proof. Let G be a finite abelian subgroup of a division algebra D. D is a $Z(D)$-Algebra. We focus on the $Z(D)$-linear span of G in D. By using proposition 1 we obtain that this span is a finite-dimensional unital $Z(D)$-division algebra. G is commutative, and hence $\langle G \rangle_{Z(D)}$ is a field and G a finite subgroup of its groups of units. The proof is finished by using theorem 1.⋄

Theorem 4 *(Herstein) Every finite subgroup of a division algebra in positive characteristics is cyclic.*

Proof. Let D be a division algebra, P the central prime subfield isomorphic to $GF(p)$ and G a finite subgroup of $E(D)$. We focus on the unital P-subalgebra $\langle G \rangle_P$ of the P-division algebra D. This division algebra is by the finiteness of G finite-dimensional. Therefor proposition 1 implies that it is a division algebra over P. P is finite and we conclude that this division algebra is finite, too. By usage of theorem 2 of Wedderburn it is a field. The corresponding theorem 1 for fields implies that G is – as a finite subgroup – cyclic.⋄

---

research library, so that he was unable to keep abreast of the mathematical literature. In 1938, he returned to Oslo to assume the Professorship of Mathematics at the university. There he taught the graduate courses in algebra and number theory, and only occasionally on mathematical logic. Skolem's Ph.D. student Øystein Ore went on to a career in the USA. Skolem served as president of the Norwegian Mathematical Society, and edited the Norsk Matematisk Tidsskrift (The Norwegian Mathematical Journal) for many years. He was also the founding editor of Mathematica Scandinavica. After his 1957 retirement, he made several trips to the United States, speaking and teaching at universities there. He remained intellectually active until his sudden and unexpected death. For more on Skolem's academic life, see Fenstad (1970).

Remark 1 The previous theorem 4 is wrong in characteristic zero. In the real quaternion algebra the quaternion group of order 8 is a finite but non-cyclic subgroup of the group of units.⋄

Herstein and Amitsur[2] have classified the finite subgroups of division algebras. A first results deals with so-called meta-cyclic groups. These groups are characterized by possessing a cyclic normal subgroup whose factor group is cyclic, too. A group having only cyclic Sylow subgroups is called a Z-group. It can be proven that Z-groups are meta-cyclic.

Theorem 5 *(Herstein) Every p-subgroup with respect to a prime number $p \neq 2$ of the group of units of a division algebra is cyclic. In particular, every subgroup of uneven order of the group of units of a division algebra is a Z-group.*

Proof. We use theorem 5.3.7 in [63] to derive that a p-group of uneven order is cyclic if it possesses exactly one subgroup of order p. This precondition is with respect to 5.3.8 in [63] valid if every abelian subgroup is cyclic. This was proven within theorem 3. The add-on follows as all Sylow subgroups are cyclic.⋄

Remark 2 The previous theorem 5 fails for $p = 2$. In the real quaternion algebra the quaternion group of order 8 is a finite but non-cyclic subgroup of the group of units. All of its subgroups are cyclic.⋄

By C_n or Z_n we denote a cyclic group of order $n \in \mathbb{N}$. If n, m are integers, then let $o_n(m) := o_{Z/nZ}(mZ)$. We formulate the classification of finite subgroups of division algebras (but we will not prove it here) in characteristic zero:

Theorem 6 *(Amitsur) Every finite subgroup of the group of units of a division algebra in characteristic zero is isomorphic one of the following groups:*

(i) C_n

---

[2]Shimshon Avraham Amitsur (born August 26, 1921, died September 5, 1994) was an Israeli mathematician. He is best known for his work in ring theory, in particular PI rings, an area of abstract algebra. Amitsur was born in Jerusalem and studied at the Hebrew University under the supervision of Jacob Levitzki. His studies were repeatedly interrupted, first by World War II and then by the Israel's War of Independence. He received his M.Sc. degree in 1946, and his Ph.D. in 1950. Later, for his joint work with Levitzki, he received the first Israel Prize in Exact Sciences. He worked at the Hebrew University until his retirement in 1989. Amitsur was a visiting scholar at the Institute for Advanced Study from 1952 to 1954. He was an Invited Speaker at the ICM in 1970 in Nice. He was a member of the Israel Academy of Sciences, where he was the Head for Experimental Science Section. He was one of the founding editors of the Israel Journal of Mathematics, and the mathematical editor of the Hebrew Encyclopedia. Amitsur received a number of awards, including the honorary doctorate from Ben-Gurion University in 1990. His students included Avinoam Mann, Amitai Regev, Eliyahu Rips and Aner Shalev.

(ii) A Z-group of the form $C_m \rtimes C_4$ for which C_4 acts per inversion on C_m and m is uneven.

(iii) A Z-group of the form $T_0 \times \cdots \times T_s$ in which the orders of these factors are pairwise prime to each other, T_0 is cyclic, every T_i, $i \in \underline{s}$ is non-cyclic of the form $C_{p^a} \rtimes (C_{q_1^{b_1}} \times \cdots \times C_{q_r^{b_r}})$, the prime numbers p, q_i, $i \in \underline{r}$ are distinct, for every $i \in \underline{r}$ the semidirect product $C_{p^a} \rtimes C_{q_i^{b_i}}$ is non-cyclic and is satisfying the following condition: if C_{p^c} is the kernel of the operation of $C_{q_i^{b_i}}$ on C_{p^a}, then one of the following cases are valid:
($q_i = 2$, $p \equiv -1 \bmod 4$, $c = 1$) or
($q_i = 2$, $p \equiv -1 \bmod 4$, 2^{c+1} does not divide $p^2 - 1$) or
($q_i = 2$, $p \equiv 1 \bmod 4$, 2^{c+1} does not divide $p - 1$) or
($q_i > 2$, q_i^{c+1} does not divide $p - 1$.)
In addition, for every non-cyclic factor $C_{p^a} \rtimes C_{q_i^{b_i}}$ within every factor T_j the statement $q_i \cdot o_{p^c}(p)$ does not divide $o_{|T/T_i|}(p)$ is valid.

(iv) $C_m \rtimes Q_{2^t}$ in which m is uneven, an element of Q_{2^t} of order 2^{t-1} centralizes the group C_m and an element of order 4 of Q_{2^t} inverts the group C_m.

(v) $Q_8 \times Z$ in which Z is a Z-group of order m presented in (i), (ii) and 2 has uneven order in $\mathbb{Z}/\mathbb{Z}m$.

(vi) $SL(2,3) \times Z$ in which Z is a Z-group of order m presented in (i), (ii) and 2 has uneven order in $\mathbb{Z}/\mathbb{Z}m$.

(vii) The binary octahedral group of order 48.

(viii) The binary icosahedral group of order 120.

Proof. see Amitsur [1], Herstein [20] and Lam [38] ⋄

2.3 Exercises

Excercise 1 *Read the article [1] of Amitsur. Determine for all finite subgroups of division algebras suitable division algebra in which they are appearing!*

Excercise 2 *Define meta-abelian and supersolvable groups by a research in the literature.*

Excercise 3 *Prove or disprove the following statements:*

(i) Every cyclic group is meta-cyclic.

(ii) The converse of (i) is valid.

(iii) Every abelian group is meta-cyclic.

(iv) The converse of (i) is valid.

(v) Direct products of meta-cyclic groups are meta-cyclic.

(vi) Semidirect products of meta-cyclic groups are meta-cyclic.

(vii) Every meta-cyclic group is supersolvable.

(viii) Every meta-cyclic group is meta-abelian.

(ix) A group for which all Sylow subgroups are cyclic is meta-cyclic.

(x) A group of squarefree order is meta-cyclic.

(xi) Dihedral groups are meta-cyclic.

(xii) Quaternion groups are meta-cyclic.

(xiii) Semidihedral groups are meta-cyclic.

Excercise 4 *Prove the following statements: An unitary algebra is an algebra with a unit element. A unital subalgebra of an algebra A is a subalgebra containing the unit element of A. A unital subalgebra is unitary. A unitary subalgebra is a subalgebra which is unitary as an algebra. A unitary subalgebra is not unital in general. (Tip: idempotent elements)*

Excercise 5 *By using an article [20] of Herstein prove the following statements (p prime number, D a skew field and U a subgroup of $E(D)$):*

(i) *If U is of order p or p^2, then U is cyclic.*

(ii) *If $p \neq 2$ and U is a p-group, then U is cyclic.*

(iii) *Is part (ii) true for $p = 2$?*

(iv) *If the order of U is uneven, then U is meta-cyclic.*

Excercise 6 *True or false: The unit group of an infinite field is cyclic. Is it possible to characterize finite fields by characteristics of their unit group?*

Excercise 7 *Determine all finite subgroups of the multiplicative group of complex numbers! How many non-isomorphic subgroups of order n are existing? Visualize them for $n \in \underline{8}$ on the complex plane!*

Excercise 8 *Are there finite subgroups in the additive group of complex numbers? On what terms do finite subgroups of the additive group of a field exist which are non-trivial? What is the answer for the multiplicative group?*

Excercise 9 *The additive group of a field is not isomorphic to the multiplicative group of a field.*

Excercise 10 *Focus on the exp-function from $\mathbb{R}$ to $\mathbb{R}_{>0}$. Is it a homomorphism for the addition and multiplication?*

Excercise 11 *Is the exponential map related to the complex numbers a homomorphism for the additive and multiplicative structure?*

Excercise 12 *Every finite generated subgroup of the multiplicative group of rational numbers is cyclic.*

Excercise 13 *Determine the finite subgroups of the multiplicative group of the real numbers.*

Excercise 14 *Determine the finite subgroups of the additive group of the real numbers.*

Excercise 15 *Are there infinite subgroups of the additive and multiplicative group of integers, rational, real and complex numbers?*

Excercise 16 *By a research of the literature determine all finite subgroups of the additive and multiplicative group of integers, rational, real and complex numbers. Are there subgroups which are simultaneously subgroups of the additive and multiplicative group of these structures?*

Excercise 17 *True or false: Every finite subgroup of a division algebra is cyclic! Analyze this question for the additive group of a division algebra.*

Excercise 18 *By a research in the literature determine all finite subgroups of the real quaternion algebra.*

Excercise 19 *The additive factor group $\mathbb{Q}/\mathbb{Z}$ is isomorphic to the group of complex roots of unity.*

Excercise 20 *The additive factor group $\mathbb{R}/\mathbb{Z}$ is isomorphic to the group of complex roots of unity of absolute value 1.*

Excercise 21 *Determine the finite subgroups of the group of units of a quaternion algebra in characteristic 2.*

Excercise 22 *Determine the finite subgroups of the group of units of a quaternion algebra in characteristic $p \geq 3$.*

Excercise 23 *Determine the finite abelian subgroups of the group of units of a quaternion algebra in arbitrary characteristic.*

Excercise 24 *True or false: Every finite subgroup of the group of units of a quaternion algebra in arbitrary characteristic is cyclic or abelian!*

Chapter 3

Normal and subnormal subgroups in unit groups of division algebras

In this chapter we summarize some results of normal and subnormal subgroups in unit groups of division algebras. For some of them we provide a proof, for the others we reference the corresponding literature. We will use some of these results in the next chapters. Therefor these results provide the reader a deeper insight for understanding these results. In addition, this chapter is included on personal interest of the author for the proofs of these results.

3.1 The theorem of Cartan-Brauer-Hua

For two elements a, b of a group we denote by $[a, b]$ the commutator of a and b. The theorem of Cartan-Brauer-Hua focus on the structure of the normal subgroups in unit groups of division algebras. Its proof is based on the so-called Hua-identity:

Proposition 3 *(Hua-identity[1]) Let D be a division algebra, $a, b \in E(D)$ with $a \neq 1$. The following statements are valid:*

(i) $b^{-1}(a-1)b = b^{-1}ab - 1$

(ii) $b^{-1}ab = a[a,b]$

(iii) $(a-1)[a-1,b] = a[a,b] - 1$

(iv) $a([a,b] - [a-1,b]) = 1 - [a-1,b]$, *and* $[a,b] - [a-1,b]$ *is not zero if* $[a,b] \neq 1$ *is valid.*

By using the identity in part (iv) we derive that every non-central element is contained in the sub-(division)algebra of D generated by all commutators of $E(D)$:

$$a = (1 - [a-1,b])(a([a,b] - [a-1,b]))^{-1}.$$

Proof. The proof is an exercise for the reader (see exercise 26). ◇

Theorem 7 *(Cartan-Brauer-Hua) Let D be a division algebra and T an unital divisions subalgebra of D. If $E(T)$ is a normal subgroup $E(D)$, then T is central or identical to D.*

Add-on: If N is a normal subgroup of $E(D)$, then N is central or the subalgebra generated by N $(= \langle N \rangle_K)$ is exactly D.

Proof. (Sysak [67]) We assume that T is not central. Let b be an element of $T \setminus Z(D)$. We focus on an element $a \in D$ such that $ab \neq ba$ is valid. $E(T)$ is a normal subgroup of $E(D)$. Therefor we conclude:

$$1 \neq [a,b] = (a^{-1}b^{-1}a)b \in T$$

and with an analog argument

$$1 \neq [a-1,b] \in T.$$

By usage of proposition 3 we derive the identity

---

[1] Hua Luogeng, or Hua Loo-Keng (born 12 November 1910; died 12 June 1985), was a Chinese mathematician famous for his important contributions to number theory and for his role as the leader of mathematics research and education in the People's Republic of China. He was largely responsible for identifying and nurturing the renowned mathematician Chen Jingrun who proved Chen's theorem, the best known result on the Goldbach conjecture. In addition, Hua's later work on mathematical optimization and operations research made an enormous impact on China's economy. Hua did not receive a formal university education. Although awarded several honorary PhDs, he never got a formal degree from any university. In fact, his formal education only consisted of six years of primary school and three years of middle school. For that reason, Xiong Qinglai, after reading one of Hua's early papers, was amazed by Hua's mathematical talent, and in 1931 Xiong invited him to study mathematics at Tsinghua University.

$$a = (1 - [a-1,b])(a([a,b] - [a-1,b]))^{-1} \in T.$$

Every element of D not commuting with b is contained in T. Let c be an element commutating with b. Then $a + c$ is not commutating with b (because a would be commutating with b). Hence $a+c$ and a are contained in T and therefor their difference c, too. We conclude the identity $T = D$. The add-on is a direct consequence of this result.⋄

3.2 The theorem of Scott

The theorem of Scott analyzes on what terms the group of unit of a division algebra is solvable.

Theorem 8 *(Scott) Let D be a division algebra. The factor group $E(D)/E(Z(D))$ possesses no non-trivial non-abelian normal subgroup. In addition, $E(D)$ is solvable if and only if K is a field.*

Proof. (Sysak [67]) We assume there is a non-central normal subgroup A of $E(D)$ such that $A/Z(D)$ is abelian. The subalgebra of D generated by A fulfills the preconditions of the theorem of Cartan-Brauer-Hua 7, and we conclude that it is identical to D. Hence A is non-abelian and two elements $a, b \in A$ exist with $1 \neq [a,b] =: c \in Z(E(D))$. We prove that $Z(E(D))$ is a subgroup of order 2 of $E(D)$ and $a^2, b^2 \in Z(E(D))$ is valid. The identities

$$\begin{aligned} [1+a, b] &= [1+a]^{-1} b^{-1} (1+a) b \\ &= (1+a)^{-1}(1+ac) \end{aligned}$$

are valid, and hence we conclude

$$\begin{aligned} [[1+a,b]b] &= [(1+a)^{-1}(1+ac), b] \\ &= (1+ac)^{-1}(1+a)(1+ac)^{-1}(1+ac^2) \\ &= (1+a)(1+ac)^{-2}(1+ac^2). \end{aligned}$$

With a similar argument we derive:

$$\begin{aligned} &[[[1+a,b],b],b] \\ &= (1+a)^{-1}(1+ac)^3(1+ac^2)^{-3}(1+ac^3). \end{aligned}$$

Because of $[1+a,b] \in A$ we get $[[[1+a,b],b],b] = 1$. Hence

$$\begin{aligned} (1+ac)^3(1+ac^3) &= (1+a)(1+ac^2)^3 \\ &= 0 \end{aligned}$$

is valid. We conclude
$$a(3c + c^3) + a^3(c^3 + 3c^5)$$
$$= a(1 + 3c^2) + a^3(3c^4 + c^6)$$
which is equivalent to
$$(1-c)^3 + a^2c^3(-1+c)^3 = 0.$$
By using $1 - c \neq 0$ we derive
$$a^2 = c^{-3} \in Z(E(D)).$$
A symmetric argument let us deduce $b^2 \in Z(E(D))$. For the element $z \in Z(E(D))$ the identity $(az)^2 = c^{-3}$ is valid because of $az \in A$ and $[az, b] = c$. Hence $z^2 = 1$ is true and we conclude that the exponent and the order of $Z(E(D))$ is 2. $Z(D)$ is the prime field of order 3 in D and a, b elements of finite order. The subgroup generated by $\{a, b\}$ is nilpotent and hence – as a torsion group – finite. By using theorem 4 the subgroup is cyclic. This is a contradiction to the choice of a, b. ⋄

3.3 The theorem of Stuth

The next result – which is formulated but not proven in details now – generalizes the previous results of Cartan-Brauer[2]-Hua and of Scott to the structure of the subnormal subgroups and was proven by Stuth in [66]:

---

[2]Richard Dagobert Brauer (born February 10, 1901, died April 17, 1977) was a leading German and American mathematician. He worked mainly in abstract algebra, but made important contributions to number theory. He was the founder of modular representation theory. Alfred Brauer was Richards brother and seven years older. Alfred and Richard were both interested in science and mathematics, but Alfred was injured in combat in World War I. As a boy, Richard dreamt of becoming an inventor, and in February 1919 enrolled in Technische Hochschule Berlin-Charlottenburg. He soon transferred to University of Berlin. Except for the summer of 1920 when he studied at University of Freiburg, he studied in Berlin, being awarded his Ph.D. on 16 March 1926. Issai Schur conducted a seminar and posed a problem in 1921 that Alfred and Richard worked on together, and published a result. The problem also was solved by Heinz Hopf at the same time. Richard wrote his thesis under Schur, providing an algebraic approach to irreducible, continuous, finite-dimensional representations of real orthogonal (rotation) groups. Ilse Karger also studied mathematics at the University of Berlin. She and Richard were married 17 September 1925. Their boys George Ulrich (1927) and Fred Gunther (1932) also became mathematicians. Brauer began his teaching career in Königsberg (now Kaliningrad) working as Konrad Knopp's assistant. Brauer expounded central division algebras over a perfect field while in Königsberg: the isomorphism classes of such algebras form the elements of the Brauer group he introduced. When the Nazi Party took over in 1933, the Emergency Committee in Aid of Displaced Foreign Scholars took action to help Brauer and other Jewish scientists. Brauer was offered an assistant professorship at University of Kentucky. Richard accepted the offer, and by the end of 1933 he was in Lexington, Kentucky, teaching in English. Ilse followed the next year with George and Fred; brother Alfred made it to the USA in 1939, but their sister Alice was killed in The Holocaust.

Theorem 9 *(Stuth) Let D be a division algebra. The following statements are valid:*

(i) *The central subgroups of $E(D)$ are exactly the solvable subnormal subgroups of $E(D)$.*

(ii) *For every non-central subnormal subgroup S of $E(D)$ the identity $C_{E(D)}(S) = Z(E(D))$ is valid.*

(iii) *The intersection of two non-central subnormal subgroups of $E(D)$ is a non-central subnormal subgroup of $E(D)$, too. In particular, there is a smallest non-central subnormal subgroup of $E(D)$ which is non-solvable.*⋄

Hermann Weyl invited Richard to assist him at Princetons Institute for Advanced Study in 1934. Richard and Nathan Jacobson edited Weyl's lectures Structure and Representation of Continuous Groups. Through the influence of Emmy Noether, Richard was invited to University of Toronto to take up a faculty position. With his graduate student Cecil J. Nesbitt he developed modular representation theory, published in 1937. Robert Steinberg, and Stephen Arthur Jennings were also Brauer's students in Toronto. Brauer also conducted international research with Tadasi Nakayama on representations of algebras. In 1941 University of Wisconsin hosted visiting professor Brauer. The following year he visited the Institute for Advanced Study and Bloomington, Indiana where Emil Artin was teaching. In 1948 Richard and Ilse moved to Ann Arbor, Michigan where he and Robert M. Thrall contributed to the program in modern algebra at University of Michigan. With his graduate student K. A. Fowler, Brauer proved the Brauer-Fowler theorem. Donald John Lewis was another of his students at UM. In 1952 Brauer joined the faculty of Harvard University. Before retiring in 1971 he taught aspiring mathematicians such as Donald Passman and I. Martin Isaacs. The Brauers frequently traveled to see their friends such as Reinhold Baer, Werner Wolfgang Rogosinski, and Carl Ludwig Siegel.

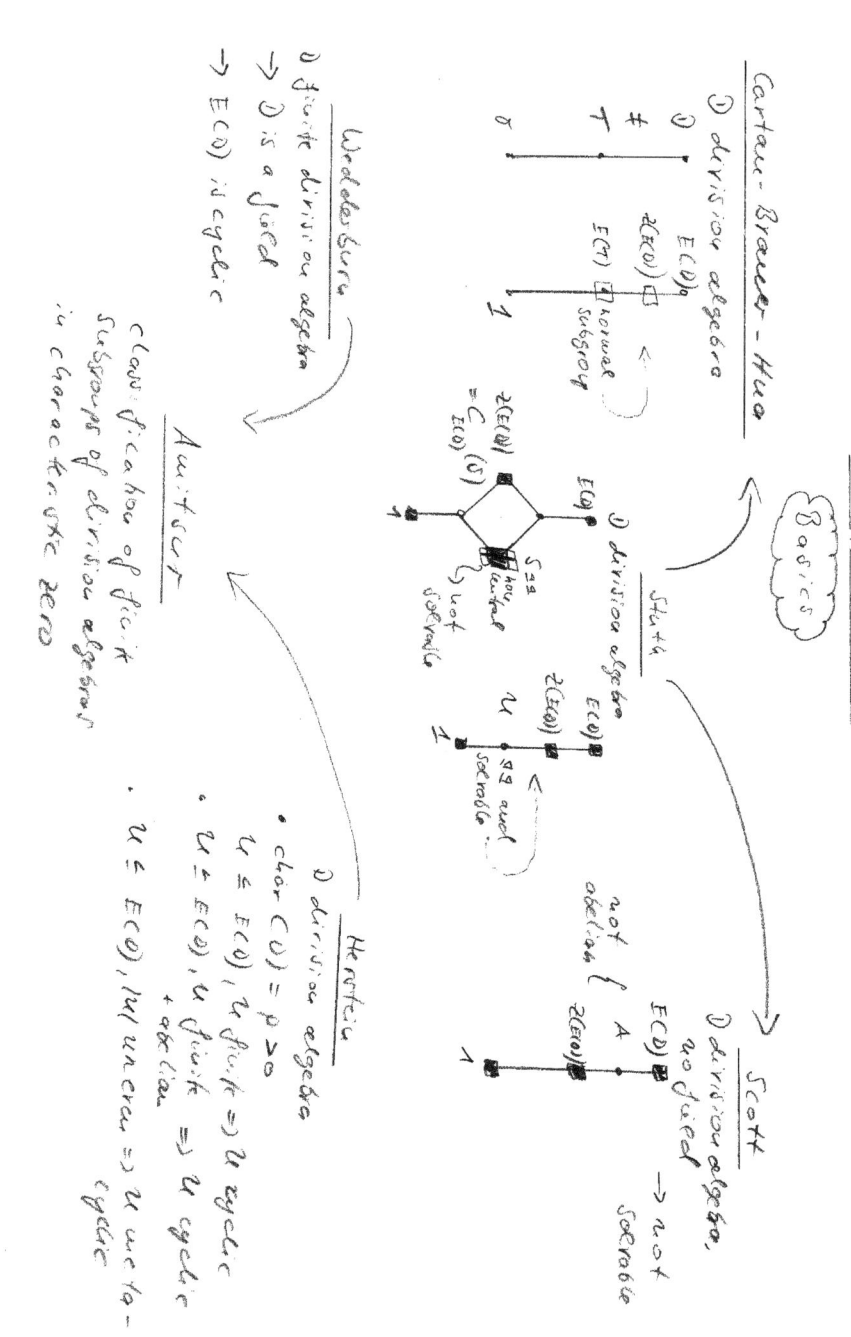

3.4 Exercises

Excercise 25 *Explain on what terms the theorem of Stuth (theorem 9) is a generalization of the theorems of Cartan-Brauer-Hua and Scott.*

Excercise 26 *Prove all identities and statements within proposition 3!*

Excercise 27 *Study the article [67] with respect to the following topics: definition of periodic groups, periodicity of the additive and multiplicative group of $\mathbb{Q}$, $\mathbb{R}$ and $\mathbb{C}$, proof of the lemmas of Herstein and the theorem of Jacobson that a division algebra with periodic group of unit is commutative. In what way is this theorem a generalization of another theorem of Jacobson?*

Excercise 28 *On what terms is the unit group of a division algebra abelian?*

Excercise 29 *On what terms is the unit group of a division algebra nilpotent? (Tip: Theorem of Scott)*

Excercise 30 *On what terms is the unit group of a division algebra solvable? (Tip: Theorem of Scott)*

Excercise 31 *On what terms is the unit group of a division algebra simple? (Tip: center and theorem of Wedderburn)*

Excercise 32 *On what terms is the unit group of a division algebra cyclic? (Tip: exercise 6)*

Excercise 33 *Let D be a division algebra. What do we know about finite normal subgroups of $E(D)$?*

Excercise 34 *Let D be a division algebra. What do we know about finite subnormal subgroups of $E(D)$?*

Excercise 35 *Do a research in literature and prove that a nilpotent and finite-generated torsion group is finite. For what reason do we need this result in this chapter? (Tip: Use the main theorem on finite-generated abelian groups and apply it on the factor groups for the nilpotent groups with respect to a central chain of finite length!)*

Excercise 36 *By using the tip in exercise 35 extend the results – if possible – to solvable groups. Is this statement also valid for non-solvable groups?*

Chapter 4

Nilradicals of Lie algebras associated to associative algebras

For every associative algebra the so-called associated Lie algebra can be derived in a natural way. In this chapter we focus on the determination of the nilradical – the greatest nilpotent ideal – of the associated Lie algebra. We describe it by the structure of the associative algebra: it is the sum the center and the nilradical of the associative algebra. In particular, the nilradical of the associated Lie algebra is an unital associative subalgebra. We assume uneven characteristic of the ground field and the separability of the factor algebra by the nilradical of the associative algebra. By assuming this we can use the well-known theorem of Wedderburn-Malcev for the existence and conjugacy of radical complements in the associative algebra.

Our strategy is to start with solvable associative algebras. For this, the generalized Jordan decomposition is the main tool for our analysis. We calculate the nilradical for our standard examples (Solomon algebras, Solomon-Tits algebras, group algebras and triangular matrices). In a next step we transfer results of Herstein[1] about simple rings and its associated Lie ring to simple

---

[1] Israel Nathan Herstein (born March 28, 1923, died February 9, 1988) was a mathematician, appointed as professor at the University of Chicago in 1951. He worked on a variety of areas of algebra, including ring theory, with over 100 research papers and over a dozen books. Herstein was born in Lublin, Poland, in 1923. His family emigrated to Canada in 1926, and he grew up in a harsh and underprivileged environment where, according to him, you either became a gangster or a college professor. During his school years he played football, ice hockey, golf, tennis, and pool. He also worked as a steeplejack and as a barker at a fair. He received his B.S. degree from the University of Manitoba and his M.A. from the University of Toronto. He received his Ph.D from Indiana University in 1948. His advisor was Max Zorn. He held positions at the University of Kansas, Ohio State University, University of Pennsylvania, and Cornell University before permanently settling at the University of Chicago in 1962. He was a Guggenheim Fellow for the academic year 1960/1961. He is known for his lucid style of writing, as exemplified by the

and semisimple algebras. The general case is studied in a last step using the results for solvable and semisimple associative algebras.

We finish this chapter by applying the results to tensor products, adjunction of a unit, matrix algebras, subalgebras etc. For this, we focus on describing the nilradical by the underlying structure, e.g. by the factors of the tensor product.

We use the term Lie nilpotent for the associated Lie algebra to pronounce that the associated Lie algebra is nilpotent. In a similar matter we use the term Lie solvable, Lie nilradical, Lie subalgebra, Lie ideal etc.

4.1 Solvable algebras

In this section we focus on solvable associative algebras and begin with the definition of the most important tool for our analysis: the generalized Jordan decomposition.

4.1.1 Generalized Jordan decomposition

Definitions and remarks 1 Let L be a K-Lie algebra and $l \in L$. We define the multiplication with l – also called adjoint representation – by $ad(l) : L \longrightarrow L, x \mapsto xl$. The nilradical of L is the greatest nilpotent ideal of L (if it is existing). The associated Lie algebra of an associative algebra A with respect to the operation $a \circ b := ab - ba$ for all $a, b \in A$ will be symbolized by A°. If S, T are subsets of A°, then we denote by $S \circ T$ the K-linear span of the set $\{s \circ t \mid s \in S, t \in T\}$. Nilpotency and solvability is used for groups and Lie algebras as usually.

By λ resp. ρ we denote the left- resp. right-regular representation of an associative algebra A. Based on definition 5.2.1 in [74] we call a polynomial fully separable if it is squarefree and separable. If A is an associative unitary K-algebra, then an element $a \in A$ is fully separable if and only if its minimal polynomial $min_{a,K}$ is fully separable. This is equivalent to the statement that $min_{a,K}$ has no multiple roots in $K[t]$ for every splitting field F of K. Another description of this property is that $min_{a,K}$ and its formal derivation $min'_{a,K}$ have no non-trivial common divisor: $gcd(min_{a,K}, min'_{a,K}) = 1$.

---

classic and widely influential Topics in Algebra, an undergraduate introduction to abstract algebra that was published in 1964, which dominated the field for 20 years. A more advanced classic text is his Noncommutative Rings in the Carus Mathematical Monographs series. His primary interest was in noncommutative ring theory, but he also wrote papers on finite groups, linear algebra, and mathematical economics. He had 30 Ph.D. students, traveled and lectured widely, and spoke Italian, Hebrew, Polish, and Portuguese. He died from cancer in Chicago, Illinois, in 1988. His doctoral students include Miriam Cohen, Susan Montgomery, Karen Parshall, Claudio Procesi, Lance Small, and Murray Schacher.

It is straightforward to prove that a, $a\rho$ and $a\lambda$ have the same minimal polynomial. Thus a is nilpotent resp. fully separable if and only if $a\rho$ or $a\lambda$ has this property. Based on definition 5.1.4.1 in [74] a pair $(r;s) \in A \times A$ is called a generalized Jordan decomposition of $a \in A$ if $a = r + s$ is valid, r and s commute, r is nilpotent and s is fully separable. The nilradical of A is denoted by $rad(A)$. A is solvable if $A/rad(A)$ – the factor algebra by its nilradical – is commutative. If $rad(A)$ possesses an complement T which is a subalgebra, then we call T a radical complement: $A = rad(A) \oplus T$. ⋄

Lemma 1 *Let K be a field, A an associative unitary finite-dimensional K-algebra and $a, r, s \in A$. If $(r;s)$ is a generalized Jordan decomposition of a, then $(ad(r); ad(s))$ is one of $ad(a)$.*

Proof. <u>Step 1</u>: It is straightforward to prove that a, $\lambda(a)$ and $\rho(a)$ possess the same minimal polynomial and $ad(a) = ad(r+s) = ad(r) + ad(s)$ is valid.

<u>Step 2</u>: We prove now that $ad(r)$ and $ad(s)$ commute: A is associative. We conclude that for all $x, y \in A$ the functions $\lambda(x)$ and $\rho(y)$ commute. Because of $rs = sr$ the functions $\lambda(r)$ and $\lambda(s)$ as well as $\rho(r)$ and $\rho(s)$ commute. We conclude:

$$\begin{aligned} ad(r)ad(s) &\\ &= (\rho(r) - \lambda(r))(\rho(s) - \lambda(s)) \\ &= \rho(r)\rho(s) - \rho(r)\lambda(s) - \lambda(r)\rho(s) + \lambda(r)\lambda(s) \\ &= \rho(s)\rho(r) - \rho(s)\lambda(r) - \lambda(s)\rho(r) + \lambda(s)\lambda(r) \\ &= (\rho(s) - \lambda(s))(\rho(r) - \lambda(r)) \\ &= ad(s)ad(r). \end{aligned}$$

<u>Step 3</u>: By usage of *Step 1* we derive that r and hence also $\rho(r)$ and $\lambda(r)$ are nilpotent. A is associative and therefor $\lambda(r)\,\rho(r) = \rho(r)\,\lambda(r)$. As a consequence the subalgebra generated by $\{\lambda(r), \rho(r)\}$ is commutative and by proposition 5 in [74] even nilpotent. Thus $ad(r) = \rho(r) - \lambda(r)$ is nilpotent, too.

<u>Step 4</u>: By usage of *Step 1* we derive that r and hence also $\rho(r)$ and $\lambda(r)$ are fully separable. A is associative and therefor $\lambda(r)\,\rho(r) = \rho(r)\,\lambda(r)$. As a consequence the subalgebra generated by $\{\lambda(r), \rho(r)\}$ is commutative and by proposition 5 in [74] separable. Thus $ad(r) = \rho(r) - \lambda(r)$ is fully separable, too. ⋄

4.1.2 The nilradical

In this section we determine the nilradical of a Lie algebra associated to a solvable associative algebra. For this we will use our results about the Jordan decomposition.

Definitions and remarks 2 An associative finite-dimensional unitary commutative algebra with separable factor algebra by the nilradical possesses exactly one radical complement: the set of all fully separable elements. If the algebra is not commutative but solvable and T is a radical complement, then $T \cap Z(A)$ is a radical complement of the center of A. This set coincides with the intersection of all radical complements of $rad(A)$ in A (see e.g. [74], chapter 5).
Let L be a Lie algebra. A subalgebra M of L is called maximal nilpotent if M is not contained in a proper nilpotent subalgebra of L.⋄

Theorem 10 *Let K be a field, A an associative unitary finite-dimensional solvable K-algebra with separable factor algebra by the nilradical and Z the radical complement of the center of A. The nilradical of $A°$ is $rad(A) \oplus Z = rad(A) + Z(A)$. In particular, the nilradical of $A°$ is an unital associative subalgebra of A and the only maximal nilpotent subalgebra of $A°$ containing $rad(A)$.*

Proof. Both ideals $rad(A)$ and Z of $A°$ are nilpotent, Hence by using a theorem of Fitting their sum is nilpotent, too. Thus the nilpotent ideal $rad(A) \oplus Z$ is contained in the nilradical of $A°$. Let N be the nilradical of $A°$ and $n \in N$. By the theorem of Wedderburn-Malcev a radical complement T exists. Let $r \in rad(A)$ and $t \in T$ such that $n = r + t$ is valid. As $rad(A)$ is contained in N we derive $t \in N$. N is a nilpotent Lie algebra and we conclude that $ad(t)$ is a nilpotent endomorphism of N. In particular, $ad(t)_{|rad(A)}$ is a nilpotent endomorphism of $rad(A)$. $A/rad(A)$ is separable and commutative and by usage of theorem 5.3.1 in [74] every element of T is fully separable. Lemma 1 implies that $ad(t)$ is fully separable, too. In particular, $ad(t)_{|rad(A)}$ is fully separable. We derive that $ad(t)_{|rad(A)}$ is nilpotent and fully separable. Lemma 1 implies $ad(t)_{|rad(A)} = 0$.[2] We conclude that t centralizes $rad(A)$. T is commutative and thus we get $t \in T \cap Z(A)$. This subalgebra is exactly Z which is proven in 5.1.4 in [74].
The identity $Z(A) = rad(Z(A)) + Z$ is valid. For an associative solvable algebra the nilradical is exactly the set of all nilpotent elements (see e.g. [4] or the section about reduced algebras in this volume). Therefor $rad(A)+Z = rad(A) + rad(Z(A)) + Z = rad(A) + Z(A)$ is valid
The add-on is a consequence of the observation that every subalgebra of $A°$ containing $rad(A)$ is – because of $A \circ A \subseteq rad(A)$ – an ideal of $A°$ and the nilradical coincide with the sum of an ideal and a subalgebra of A. ⋄

---

[2]Every element possesses at most one Jordan decomposition (see e.g. chapter 5 in [74]).

4.2 Standard examples of solvable associative algebras

4.2.1 Solvable group algebras

In this section we determine the nilradical of the associated Lie algebra $(KG)^\circ$ for a modular solvable group algebra KG with respect to dihedral and quaternion groups. We need the following general insight about solvable group algebras:

Preliminary remark 1 Let K be a field and G a finite group. By using 3.2.20 in [74] the group algebra KG is solvable if and only if G is abelian or $char(K) = p$ is valid and G' – the derived subgroup of G – is a p-group (p a prime number). If G is abelian, then $(KG)^\circ$ is nilpotent. Let $char(K) = p$ and G' a p-group. Then there exists a p-Sylow subgroup P containing the derived subgroup. Hence P is a normal p-Sylow subgroup of G, and therefor its the only p-Sylow subgroup by using Sylow's theorem. The theorem of Schur-Zassenhaus assured a complement H of P in G. If α is the linearization of the natural epimorphism from G onto the factor group G/P, then the kernel of α is well-known: $Kern\,\alpha = KGAug(KP) = Aug(KP)KG$, and $Aug(KP)$ is the augmentation ideal of KP. By a theorem of Wallace $Aug(KP)$ is nilpotent, and thus $Kern\,\alpha$ is nilpotent, too. The factor algebra KG modulo $Kern\,\alpha$ is isomorphic to $K(G/P)$ which is isomorphic to KH. KH is – using Maschke's theorem – semisimple, and we conclude by 1.9.4 in [74] that it is even separable. We derive $rad(KG) = KG\,Aug(KP)$, and KH is a separable radical complement in KG. In particular, the dimension of the nilradical is $|G| - |H|$. By using results in [74], chapter 3, the intersection of all radical complements is the radical complement of the center of KG. In particular, the radical complement of the center of KG is contained in KH. The nilradical of $(KG)^\circ$ is determined – by using theorem 10 – as the direct sum of the radical complement of the center of KG and the nilradical ($=KGAug(KP)$) of KG. By using the same theorem it is also the sum of the radical of KG with the center of KG, and the latter subalgebra is K-spanned by the conjugacy class sums of G. The radical complement of the center of KG can be constructed as follows: calculate the generalized Jordan decomposition for a basis of $Z(KG)$. The K-linear span of the calculated fully separable parts is the desired radical complement of the center of KG. This calculation is done within remark 3.⋄

Let G be a group, K a field and T a finite subset of G. By $\overline{T}$ we denote the sum of all elements of T in KG. $C(G)$ symbolizes the set of all conjugacy classes of G and $k(G)$ the number of these classes. If U is a subgroup of G, then the core of U in G is the greatest normal subgroup of G contained in U. Its exactly the intersection of all conjugates of H in G and is symbolized by $core_G(U) = \bigcap_{g \in G} U^g$. By using the preliminary remark we prove now:

Theorem 11 (Lie nilradical of solvable group algebras) *Let G be a finite group and K a field with $char(K) = p > 0$ such that KG is solvable. If H is a p'-Hall subgroup and P is the normal p-Sylow subgroup of G, then $(Aug(KP)KG) \oplus K(Z(G) \cap H)$ is the nilradical of $(KG)^\circ$. In particular, the dimension of the Lie nilradical is $|G| - |H| + |Z(G) \cap H|$.*

Proof. We use some results of theorem 16 and the preliminary remark 1. KH is a radical complement and $Z(KG) \cap KH$ is the radical complement of $Z(KG)$. We have to prove that this radical complement is exactly $K(Z(G) \cap H)$. Then the proof is finished by using theorem 10. Surely, $K(Z(G) \cap H)$ is contained in the other substructure. Let $x := \sum_{h \in H} k_h h$ an element of $Z(KG) \cap KH$. The center of KG is spanned by the conjugacy class sums as a K-space, and we present x as $\sum_{C \in \mathcal{C}(G)} k_C \overline{C}$. Let $h \in H$ with $k_h \neq 0$, then there exists a conjugacy class C of G and an element $c \in C$ with $h = c$ and $k_C \neq 0$. All conjugates of c – which is the set C – are contained in H: the coefficient of all elements of C is exactly k_C and the conjugacy classes partition G. Therefor all conjugates of c satisfy the identity $\sum_{h \in H} k_h h = \sum_{C \in \mathcal{C}(G)} k_C \overline{C}$ and are contained in H. We have proven $h^G = C \subseteq H$. By using $h^G \subseteq H$ we derive $h \in \bigcap_{g \in G} H^g$. This set is the core of H in G. This normal subgroup has trivial intersection with P, and thus every element of the core commutates with all elements of P. H is abelian and $G = PH$ is valid. This implies that the core of H in G is central in G. We have proven $h \in Z(G) \cap H$ as desired.⋄

Corollary 1 *Let p be a prime number, K a field with $char(K) = p \geq 3$, $n \in \mathbb{N}$ and G a group of order $2 \cdot p^n$. The following conditions are valid:*

(i) If a 2-Sylow subgroup is central, then $(KG)^\circ$ is nilpotent.

(ii) If a 2-Sylow subgroup is not central, then $rad(KG) \oplus K \cdot 1_G$ is the nilradical of $(KG)^\circ$ of dimension $2p^n - 1$.

Proof. The p-Sylow subgroup is a normal subgroup of order p^n and of index 2. The complement H in the preliminary remark 1 is of order 2. Thus the radical complement KH is of dimension 2, and the radical complement of the center of KG is KH or $K \cdot 1_G$ (using again the preliminary remark 1). Now the proof is finished using again the preliminary remark 1.⋄

For the dihedral groups we conclude:

Corollary 2 *Let p be a prime number, K a field with $char(K) = p \geq 3$, $n \in \mathbb{N}$ and $G = D_{2p^n}$. $rad(KG) \oplus K \cdot 1_G$ is the nilradical of $(KG)^{\circ}$ of dimension $2p^n - 1$.*⋄

For the quaternion groups we derive:

Corollary 3 *Let p be a prime number, K a field with $char(K) = p \geq 3$, $n \in \mathbb{N}_{\geq 2}$ and $G = Q_{4p^n}$. $rad(KG) \oplus KZ(G)$ is the nilradical of $(KG)^{\circ}$ of dimension $4p^n - 2$.*

Proof. Let $a, b \in G$ with $o(a) = 2p^n$, $o(b) = 4$, $z = b^2 = a^{p^n}$ such that G is generated by a, b, the center of G is generated by z and $a^b = a^{-1}z$ is valid. We use the results of the preliminary remark 1. The derived subgroup of G is generated by a^2 and is thus a p-group of order p^n. One of its complement H is generated by b and is of order 4. 1 and z are fully separable because $p \neq 2$ is valid, and H is non-central. Therefor the radical complement of the center of KG is contained in KH and possesses the dimension 2 or 3. By usage of theorem 11 its exactly $K(Z(G) \cap H)$ and we conclude that it is of dimension 2.⋄

At the end of this section we determine (as announced earlier in this section) the generalized Jordan decomposition of the conjugacy classes within the center $Z(KG)$ of KG for solvable KG:

Remark 3 *(Generalized Jordan decomposition of the conjugacy class sums)* Let K be a field with $char(K) = p \neq 0$, G a finite group with normal p-Sylow subgroup P and abelian Hall complement H. $Z(KG) = (rad(KG) \cap Z(KG)) \oplus K(Z(G) \cap H)$ is valid using theorem 11. $rad(KG)$ is K-linear spanned by the elements $(p-1)h$ with $p \neq 1, p \in P, h \in H$ (using the preliminary remark 1).

Let g^G be a conjugacy class of G for an element g which is not central in G. Thus the length of g^G is at least 2.
We begin to focus on the case $g = p \in P$. Let $C(p)$ be a set of conjugators of p in G which is one of the minimal sets M in G with $p^G = p^M$. The following equation is valid:

$$\overline{p^G} = \sum_{x \in C(p)} p^x$$
$$= \sum_{x \in C(p)} ((p^x - 1) + 1)$$
$$= (\sum_{x \in C(p)} (p^x - 1)) + \mid p^G \mid \cdot 1_{KG}.$$

The second summand which is a multiple of 1 is surely central. Thus the first summand which lies in the nilradical of KG is central, too. As a consequence we have found the decomposition. The fully separable part is zero (the class sum consequently nilpotent) if and only if the length of the conjugacy class is divided by p. For example, this is the case if P is abelian.

Let $g = h \in H$ with $h^G \subseteq H$. Following the proof of theorem 11 the element h is central in G. h^G is no subset of H. Let $C(h)$ be a set of conjugators of h in G. Then $\mid C(h) \mid = \mid h^G \mid$ is valid. For every $g \in C(h)$ elements $h_g \in H$ and $p_g \in P$ exist with $g = h_g p_g$. As H is abelian, we derive $h^g = h^{p_g}$. This element can be re-written to $(p_g^{-1} - 1)h p_g + h(p_g - 1) + h$. Thus there exists an element $j \in rad(KG)$ with $\overline{h^G} = j + \mid h^G \mid_K \cdot h$. H is abelian, and therefor H is contained in $C_G(h)$. We conclude that $\mid h^G \mid$ is a power of p or h ist central. By our assumption h is not central, and we derive that h^G is nilpotent (because of $char(K) = p$).

Let $g = ph$ with $p \neq 1 \neq h$. We begin our analysis by assuming that p is central. Thus h is not central and $\overline{(ph)^G} = \overline{h^G} \cdot p$ is valid. By our analysis so far we deduct the nilpotency of $\overline{h^G}$ and the one of $\overline{(hp)^G}$. Now let h central and p not central. We calculate $\overline{(ph)^G} = \overline{p^G} \cdot h$. The conjugates of ph in G are closely connected to the ones of p in G, because h is central. There exist elements $j \in rad(KG)$ and $k \in K$ with $\overline{h^G} = j + k \cdot 1_G$. We conclude $\overline{(ph)^G} = h \cdot j + k \cdot h$. Because $h \in Z(G) \cap H$ is valid, the element $h \cdot j$ is central in KG and contained in $rad(KG)$. In this case we have determined the Jordan decomposition. Now let p and h be not central. We proceed as in the case for h^G. Let $C(ph)$ be a set of conjugators of ph in G. For every $x \in C(ph)$ let $p_x \in P$ and $h_x \in H$ such that $x = h_x p_x$ is valid. We derive:

$$\overline{(ph)^G} = \sum_{x \in C(ph)} (p^x - 1)h^x + \sum_{x \in C(ph)} h^x.$$

The first sum is contained in $rad(KG)$. For the second sum we calculate further:

$$\sum_{x \in C(ph)} h^x = \sum_{x \in C(ph)} h^{p_x} = \sum_{x \in C(ph)} (p_x^{-1}) + h(p_x - 1) + \sum_{x \in C(ph)} h.$$

Thus there exists an element $j \in rad(KG)$ such that

$$\sum_{x \in C(ph)} (ph)^x = j + \mid C(ph) \mid_K \cdot h$$

is valid. This decomposition is unique as direct sum of elements of $rad(KG)$ and KH. $\sum_{x \in C(ph)} (hp)^x$ decomposes in $Z(KG)$ as a sum of elements of $Z(KG) \cap rad(KG)$ and of $Z(KG) \cap KH$. This sum is K-direct. Therefor both decompositions are identically. h is not central, and thus $\mid C(ph) \mid$ must be divided by p and is zero. In particular, the whole conjugacy class sum is nilpotent.

Only the case $g \in Z(G)$ is remaining now. For this we prove that $P \cap Z(G)$ is the normal p-Sylow subgroup with abelian complement $Z(G) \cap H$ in $Z(G)$. If $g \in Z(G)$, then there exist $p \in Z(G) \cap P$ and $h \in Z(G) \cap H$ such that $g = ph = (p-1)h + h$ is valid. This is the desired decomposition. The prime divisors of $\mid Z(G) \mid$ are ones of $\mid G \mid$, too. As an abelian group $Z(G)$ is the direct product of its Sylow subgroups. The theorem of Sylow implies that we can conjugate them into the corresponding ones of G. But the center is central and the conjugation on the center is the identity. As a consequence they are contained in every corresponding Sylow subgroup of G.⋄

As another application we determine within the next sections the Lie nilradical for our standard examples: the Solomon-Tits algebras, the Solomon algebras and the algebras of upper and lower triangular matrices.

4.2.2 Triangular matrices

Let K be a field and $n \in \mathbb{N}$. The algebra of lower triangular matrices are examples of solvable associative K-algebras with separable factor algebra by its nilradical. The nilradical of the algebra of lower triangular matrices of $K^{n \times n}$ - denoted by $\delta_{u,n}$ - is the subalgebra of strict lower triangular matrices - symbolized by $s\delta_{u,n}$. The nilradical has the dimension $\sum_{i=1}^{n-1} i = \frac{1}{2}(n-1)n$. $\delta_{u,n}$ is central, its center is of dimension 1. By theorem 10 we derive:

Corollary 4 *For the Lie nilradical of the algebra of lower triangular matrices the following identities are valid:*

(i) $nil(\delta_{u,n}{}^\circ) = s\delta_{u,n} \oplus K \cdot 1$

(ii) $dim_K(nil(\delta_{u,n}{}^\circ)) = 1 + \frac{1}{2}(n-1)n$.⋄

Let K be a field and $n \in \mathbb{N}$. The algebra of upper triangular matrices are examples of solvable associative K-algebras with separable factor algebra by its nilradical. The nilradical of the algebra of upper triangular matrices of $K^{n \times n}$ - denoted by $\delta_{o,n}$ - is the subalgebra of strict upper triangular matrices - symbolized by $s\delta_{o,n}$. The nilradical has the dimension $\sum_{i=1}^{n-1} i = \frac{1}{2}(n-1)n$. $\delta_{o,n}$ is central, its center is of dimension 1. By theorem 10 we derive:

Corollary 5 *For the Lie nilradical of the algebra of upper triangular matrices the following identities are valid:*

(i) $nil(\delta_{o,n}{}^\circ) = s\delta_{u,n} \oplus K \cdot 1$

(ii) $dim_K(nil(\delta_{o,n}{}^\circ)) = 1 + \frac{1}{2}(n-1)n.\diamond$

4.2.3 Solomon algebras in characteristic zero

Let K be a field of characteristic zero and $n \in \mathbb{N}$. By D_n we denote the Solomon algebra. It is the K-linear span of all class sums of so-called defect classes in KS_n: if $\alpha \in S_n$, then $D(\alpha) := \{i \mid i\alpha > (i+1)\alpha\}$. The Solomon algebra is the K-linear span of $\{\sum_{D(\alpha)=D} \alpha \mid D \subseteq \underline{n-1}\}$. The surprising insight of Luis Solomon was that products of two defect class sums is a linear combination of defect class sums. Thus the linear span of the defect class sums is indeed an associative algebra. The Solomon algebra possesses the dimension 2^{n-1}, and its factor algebra by the nilradical is of dimension $p(n)$ - the number of partitions of n. D_n is a solvable associative algebra with separable factor algebra by its nilradical. Its center is semisimple. Thus the intersection of the center with the nilradical is trivial. For even resp. uneven n the center is of dimension 3 resp. 2. For details the reader may study the dissertation of Thorsten Bauer [4], in particular chapter 3, in which the descriptions of the radical is contained, too. By theorem 10 we derive:

Corollary 6 *For the Lie nilradical of the Solomon algebra in characteristic zero the following identities are valid:*

(i) $nil(D_n{}^\circ) = rad(D_n) \oplus Z(D_n)$

(ii) $dim_K(nil(D_n{}^\circ)) = 2^{n-1} - p(n) + 3$, n *even.*

(iii) $dim_K(nil(D_n{}^\circ)) = 2^{n-1} - p(n) + 2$, n *uneven.*$\diamond$

4.2.4 Solomon-Tits algebras

Let K be a field and $n \in \mathbb{N}$. By $S(n,k)$ we denote the so-called Stirling numbers with resp. to $k \in \underline{n}_0$. Its the number of unordered set partitions of $\underline{n}$ consisting of exactly k subsets. In [76] the Solomon-Tits algebra $K\Pi_n$ is analyzed in details for several questions and problems based on a paper by Manfred Schocker. The algebra is defined as the monoid algebra $K\Pi_n$ with resp. to the monoid Π_n. This monoid consists of all ordered set partitions of $\underline{n}$. If $(P_1, \cdots, P_l)$ and $(Q_1, \cdots Q_k)$ are two elements of this monoid, then their product is defined by

$$(P_1, \cdots, P_l) \wedge_n (Q_1, \cdots, Q_k) :=$$
$$(P_1 \cap Q_1, P_1 \cap Q_2, \cdots, P_1 \cap Q_k, \cdots, P_l \cap Q_1, P_l \cap Q_2, \cdots, P_l \cap Q_k)^\emptyset.$$

The symbol $^\emptyset$ signalizes that empty sets are deleted from this tuple. $K\Pi_n$ is again an example for a solvable associative algebra with separable factor algebra by its nilradical.

The nilradical of $K\Pi_n$ is described in several ways which are not presented here (see e.g. chapter 2 in [76]), and its dimension is $dim_K(rad(K\Pi_n)) = \sum_{k=0}^{n}(k! - 1)\, S(n, k)$ (see corollary 8 in [76]). The center of $K\Pi_n$ is one-dimensional because $K\Pi_n$ is central (see theorem 13 in [76]). Therefor we can apply theorem 10 and conclude:

Corollary 7 *For the Lie nilradical of the Solomon-Tits algebra the following identities are valid:*

(i) $nil(K\Pi_n{}^\circ) = rad(K\Pi_n) \oplus K \cdot 1$

(ii) $dim_K(nil(K\Pi_n{}^\circ)) = 1 + \sum_{k=0}^{n}(k! - 1)\, S(n, k).$ ⋄

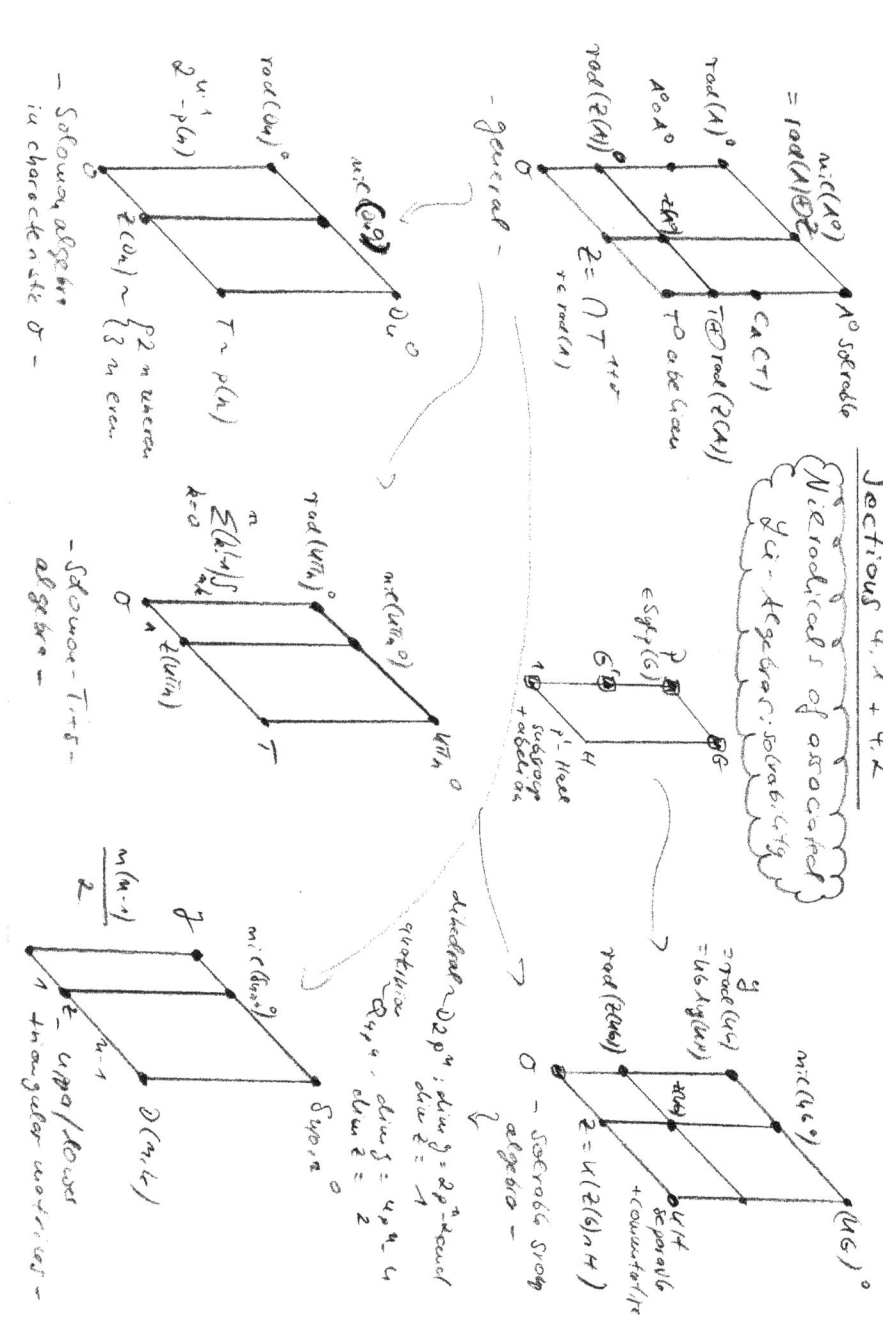

4.3 Herstein's analysis for simple associative rings

Herstein studied in several articles (see e.g. [21], [22], [23] and the articles mentioned within these articles) the influence of the associative structure of a ring on its associated structures like the Lie or Jordan ring. In his opinion this influence is wide and he proved this within the quoted articles. Of special interest was for him the case of a simple associative ring, and such rings are important for our determination of the Lie nilradical, too.

The associated Lie ring of an associative ring R is defined by the multiplication $a \circ b := ab - ba$. Associative rings are $\mathbb{Z}$-algebras. Hence the terms nilpotency, solvability, ideal, center, characteristic etc. are defined for Lie rings as for $\mathbb{Z}$-algebras. For example, if S, T are subsets of a ring, then the set $S \circ T$ is exactly the $\mathbb{Z}$-algebra span of $\{s \circ t \mid s \in S, t \in T\}$ which is the smallest Lie subring of R containing this set.

We summarize those results of Herstein which are important for our analysis. Proofs are not given here. The reader may study [22], theorem 2, corollaries 1 and 2 to get a deeper insight in this topic.

Theorem 12 *(Herstein) Let R be a simple associative ring. The following statements are valid:*

(i) If L is an ideal of R° and $char(R) \neq 2$, then L is central or $R \circ R$ is contained in L.

(ii) Either R is a field or $R = R \circ R$ is valid.

(iii) $R \circ R = (R \circ R) \circ (R \circ R).$ ⋄

As a consequence we determine the nilradical and the solvable radical of the associated Lie ring of a simple associative ring:

Theorem 13 *Let R be a simple associative ring with $char(R) \neq 2$. The following statements are valid:*

(i) The center of R is the greatest nilpotent ideal of the associated Lie ring R°.

(ii) The center of R is the greatest solvable ideal of the associated Lie ring R°.

Proof. The center is an abelian ideal of the Lie ring R°. We have to prove in part (ii) that every solvable Lie ideal is contained in the center. If R is a field, then $R = Z(R)$ is the greatest nilpotent and solvable ideal of R°. Let R be no field and L a solvable ideal of R°. We assume that L is not central. By using theorem 12 the subalgebra L contained the ideal $R \circ R$. For this ideal the same theorem implies $R = R \circ R$. Hence L is identical to R. $L = L \circ L$ is valid and L is solvable. We conclude $L = R = 0$ which is a contradiction.⋄

4.4 Simple and semisimple algebras

4.4.1 Simple algebras

In this section we transfer the results of Herstein to simple K-algebras. For this it is important that the ring possesses a unit. Recall that a K-algebra is also a ring and that every ring is a $\mathbb{Z}$-algebra.

Proposition 4 *Let A be an associative unitary K-algebra. The following conditions are valid:*

(i) *Every left ideal of the ring A is a left ideal of the K-algebra A and vice versa.*

(ii) *Every right ideal of the ring A is a right ideal of the K-algebra A and vice versa.*

(iii) *Every ideal of the ring A is an ideal of the K-algebra A and vice versa.*

(iv) *Every K-subalgebra of A is a subring of A.*

(v) *A is as a ring simple if and only if A is as a K-algebra simple.*

Proof. ad(i)-(iii): We have to prove that the substructures of interest are K-linear subspaces. For this it is important that the algebra is unitary. Let $a \in A$ and $k \in K$. Then the identity $ka = (k1_A)a = a(k1_A)$ is valid. By this identity the parts (i)-(iii) are a direct consequence. Part (iv) can be derived by (iii), and part (v) is a consequence of (iii).⋄

Example 1 The complex number field $\mathbb{C}$ is a two-dimensional associative unitary $\mathbb{R}$-algebra. The only $\mathbb{R}$-subalgebras are $\mathbb{R} \cdot x$ with $x \in \mathbb{C}$. As a ring $\mathbb{C}$ possesses several subrings which are no $\mathbb{R}$-algebras like $\mathbb{Z}$ or $\mathbb{Q}$.⋄

Remark 4 Let A be an associative, but not necessarily unitary K-algebra and L a left ideal of the ring A (The following analysis is also valid for right ideals of the ring A.). The K-linear generation $\langle L \rangle_K$ of L is a K-left ideal of the K-algebra A containing L. Its the smallest K-left ideal of A containing L. Let us focus now on the set $\hat{L} := \{l \mid l \in L \wedge Kl \subseteq L\}$. Then $\hat{L}$ is again a K-left ideal of A. It is the greatest K-left ideal of A contained in L. These three structures are identical for unitary algebra (see proposition 4), but in general they are distinct as the following example demonstrates us:

We focus on the non-unitary and nilpotent associative $\mathbb{R}$-algebra of strict lower triangular matrices in $\mathbb{R}^{3\times 3}$ and within these on the set

$$\begin{pmatrix} 0 & 0 & 0 \\ a & 0 & 0 \\ b & c & 0 \end{pmatrix},$$

for which b is a real number and a, c are integers. This set is an ideal of the corresponding ring-structure but no $\mathbb{R}$-linear space. The greatest $\mathbb{R}$-ideal contained in this set is

$$\begin{pmatrix} 0 & 0 & 0 \\ 0 & 0 & 0 \\ b & 0 & 0 \end{pmatrix},$$

and the smallest $\mathbb{R}$-ideal containing this set is the whole algebra.

Every K-algebra A can be embedded into a unitary K-algebra A^K. Some theorems are proven for unitary algebras and are transferred afterwards by using this construction to non-unitary ones. This concept fails for this problem: an ideal of the ring A is an ideal of the ring A^K if and only if it is a K-linear space.⋄

For a K-Lie algebra L the symbol $Rad(L)$ is used for the greatest solvable ideal which is called the solvable radical.

Theorem 14 *Let A be a right Artinian simple associative K-algebra with $char(K) \neq 2$. The following conditions are valid:*

(i) The center of A is exactly the nilradical of A°: $nil(A^\circ) = Z(A)$. In particular, $nil(A^\circ)$ is a unital associative K-subalgebra of A.

(ii) The center of A is exactly the solvable radical of A°: $Rad(A^\circ) = Z(A) = nil(A^\circ)$. In particular, $Rad(A^\circ)$ is a unital associative K-subalgebra of A which coincides with the nilradical of A°.

(iii) If A is central, then the identity $nil(A^\circ) = Rad(A^\circ) = K \cdot 1_A$ is valid.

Proof. A is right Artinian and simple and hence A is unitary (because it is – by using a theorem of Wedderburn – isomorphic to a matrix algebra over a division algebra). We conclude that A is as a ring simple if and only if it is simple as a K-algebra (see proposition 4). Therefor we apply theorem 13. Let L be a solvable K-ideal of A°. Then L is of course an ideal of the Lie ring A° and still solvable. The theorem is proven by using theorem 13.⋄

4.4.2 Direct products and semisimple algebras

In this section we transfer our results to direct products and semisimple algebras. On the direct product of two associative algebras the Lie multiplication is define componentwise: $(a;b) \circ (c;d) := (a \circ c; b \circ d)$.

Within the next proposition the reader may prove as an exercise that maximal nilpotent and solvable ideals of direct products are determined by direct products. So-called diagonals are not existing.

Proposition 5 *Let A, B be associative K-algebras and L an K-ideal of $(A \times B)^\circ$. We define $L_A := \{a \mid a \in A, \exists b \in B : (a;b) \in L\}$ and $L_B := \{b \mid b \in B, \exists a \in A : (a;b) \in L\}$. The following conditions are valid:*

(i) L_A is an K-ideal of A°.

(ii) L_B is an K-ideal of B°.

(iii) L is contained in $L_A \times L_B$.

(iv) If L is nilpotent, then L_A, L_B are nilpotent.

(v) If L is solvable, then L_A, L_B are solvable.

(vi) If L is maximal nilpotent, then $L = L_A \times L_B$ is valid.

(vii) $nil((A \times B)^\circ) = nil(A^\circ) \times nil(B^\circ)$

(viii) If L is maximal solvable, then $L = L_A \times L_B$ is valid.

(ix) $Rad((A \times B)^\circ) = Rad(A^\circ) \times Rad(B^\circ).\diamond$

As a consequence we derive the Lie nilradical and Lie solvable radical of semisimple associative algebras:

Theorem 15 *Let A be an associative right Artinian semisimple K-algebra, $char(K) \neq 2$, $n \in \mathbb{N}$ and $I_1, \cdots, I_n$ simple ideals of A such that A is the direct product of these ideals. The following identities are valid.:*

(i) $Z(A) = Z(I_1) \times \cdots \times Z(I_n)$

(ii) $nil(A^\circ) = nil((I_1)^\circ) \times \cdots \times nil((I_n)^\circ) = Z(A)$
In particular, $nil(A^\circ)$ is a unital and central associative K-subalgebra of A.

(iii) $Rad(A^\circ) = Rad((I_1)^\circ) \times \cdots \times Rad((I_n)^\circ) = Z(A) = nil(A^\circ)$
In particular, $Rad(A^\circ)$ is a unital and central associative K-subalgebra of A which coincides with the nilradical of A°.

Proof. A right Artinian associative semisimple K-algebra is – by using a theorem of Wedderburn-Artin[3] – unitary and a direct product of the ideals $I_1, \cdots, I_n$. Hence this theorems is an inductive consequence of proposition 5 and theorem 14. For the induction argument it is important that these ideals are right Artinian, simple and unitary.◇

---

[3]Emil Artin, born March 3, 1898, died December 20, 1962. Artin was one of the leading algebraists of the century, with an influence larger than might be guessed from the one volume of his Collected Papers edited by Serge Lang and John Tate. He worked in algebraic number theory, contributing largely to class field theory and a new construction of L-functions. He also contributed to the pure theories of rings, groups and fields. The influential treatment of abstract algebra by van der Waerden is said to derive in part from Artin's ideas, as well as those of Emmy Noether. Artin solved Hilbert's seventeenth problem in 1927. Artin was also an important expositor of Galois theory, and of the group cohomology approach to class ring theory (with John Tate), to mention two theories where his formulations became standard. In 1957, Artin wrote a book on geometric algebra, an insightful development of the classical groups in a Kleinian context. He also developed the theory of braids as a branch of algebraic topology.

Herstein

R simple ring
char $(R) \neq 2$

• —•—•——
0 $Z(R)$ R

$Z(R) = \text{nil}(R^0)$
$= \text{Rad}(R^0)$

• every non-central ideal of R^0 contains $R \circ R$
• $R \circ R = (R \circ R) \circ (R \circ R)$
• R no Jord ideal
$\Rightarrow R = R \circ R$

Sections 4.3 + 4.4

Jacobson for simple and semisimple algebras

•———•———•
0   $Z(A)$ A^0

A^0 simple algebra
char $(A) \neq 2$

$Z(A) = \text{rad}(A^0)$
$= \text{Rad}(A^0)$

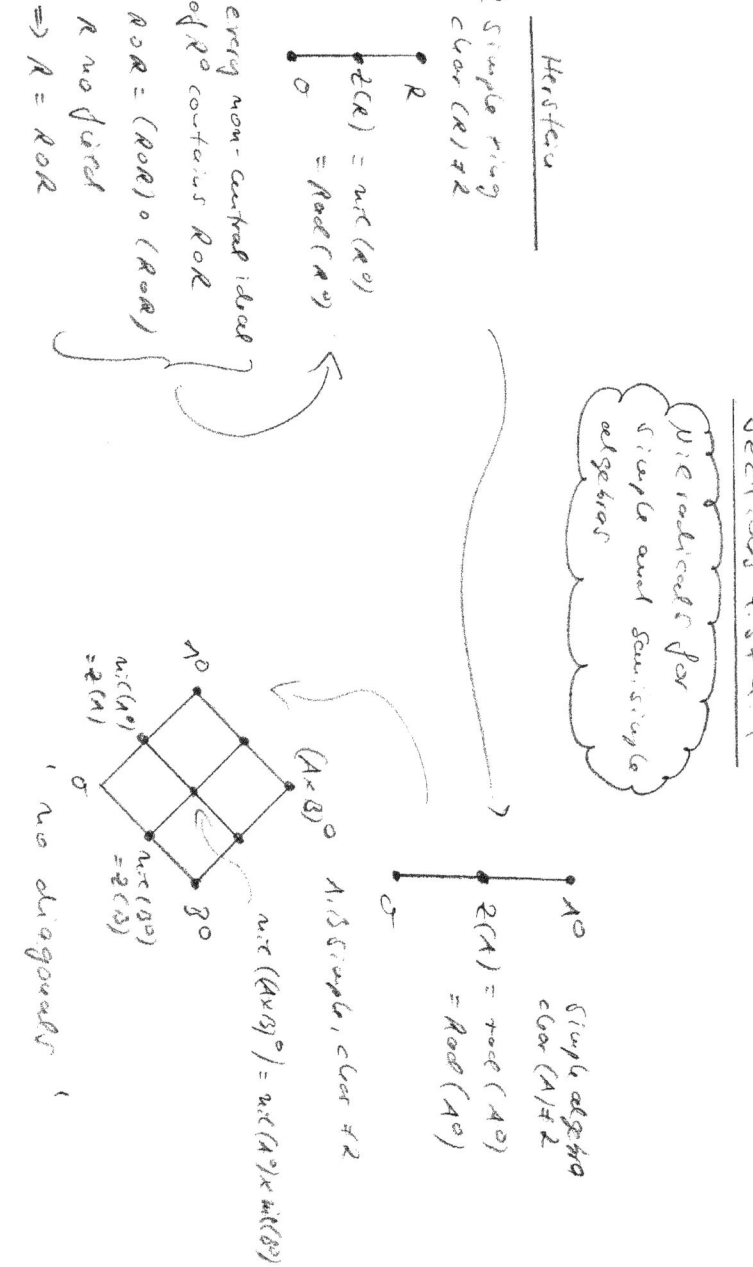

A, B simple, char $\neq 2$

$\text{nil}((A \times B)^0) = \text{nil}(A^0) \times \text{nil}(B^0)$

$(A \times B)^0$

A^0 B^0

$\text{nil}(A^0)$ $\text{nil}(B^0)$
$= Z(A)$ $= Z(B)$

0

• no diagonals •

4.5 Associative algebras with separable factor algebra by its nilradical

In this section we use the results for solvable and semisimple associative algebras to determine the Lie nilradical for associative algebras with separable factor algebra by their associative nilradical.

Proposition 6 *Let K be a field, A a finite-dimensional associative unitary K-algebra with separable factor algebra by its nilradical and S an unital K-subalgebra of A containing $rad(A)$ such that $S/rad(A)$ is central in $A/rad(A)$. Then $rad(S) = rad(A)$ is valid and $S/rad(S)$ is separable.*

Proof. A characterization of the separable K-algebra $A/rad(A)$ proven in [49] is that $A/rad(A)$ is semisimple and the centers of the simple ideals $I_1, \cdots, I_n$ decomposing $A/rad(A)$ are separable field extensions of K. $S/rad(A)$ is a subalgebra of the center of $A/rad(A)$ which is $Z(I_1) \times \cdots \times Z(I_n)$. By results in [74] a commutative K-algebra is separable if and only if every element is fully separable. Hence unital subalgebras are again separable for those algebras. In particular, the commutative unital K-subalgebra $S/rad(A)$ of the center of $A/rad(A)$ is separable. A consequence of this is that the condition $rad(S) = rad(A)$ is valid.⋄

Theorem 16 *Let K be a field with $char(K) \neq 2$ and A a finite-dimensional associative unitary K-algebra with separable factor algebra by its nilradical. Then $rad(A) + Z(A)$ is the nilradical of A°. In particular, the nilradical of A° is an unital associative K-subalgebra of A.*

Add-on: $rad(Z(A)) = rad(A) \cap Z(A)$, $Z(A)/rad(Z(A))$ is separable, and for every radical complement T of $rad(A)$ in A is $T \cap Z(A) = (\bigcap_{r \in rad(A)} T^{1+r}) \cap Z(A)$ the radical complement of $rad(Z(A))$ in $Z(A)$. In particular, for every radical complement T of $rad(A)$ in A the identity $nil(A^\circ) = rad(A) \oplus (Z(A) \cap T)$ is valid.

Proof. By theorem 15 the center of the semisimple algebra $A/rad(A)$ is the nilradical of $(A/rad(A))^\circ$. $nil(A^\circ)/rad(A)$ is a nilpotent ideal of $(A/rad(A))^\circ$ and therefor its contained in the center of $A/rad(A)$. The fundamental homomorphism theorem implies the existence of a unital K-subalgebra S containing $rad(A)$ such that $S/rad(A) = Z(A/rad(A))$ is valid. With respect to proposition 6 the algebra $S/rad(S)$ is separable, and $rad(S) = rad(A)$ is valid. As $S/rad(S)$ is commutative and $rad(S)$ nilpotent, S is in fact a unital solvable K-subalgebra of A. By the result for solvable algebras 10 we derive the identity $nil(S^\circ) = rad(A) + Z(S)$. $nil(A^\circ)/rad(A)$ is central in $(A/rad(A))^\circ$. Thus we get $nil(A^\circ)/rad(A) \subseteq$

$nil(A°/rad(A))$, and by this $nil(A°) \subseteq S$. Therefor $nil(A°)$ is a nilpotent ideal of $S°$, and we conduct $nil(A°) \subseteq nil(S°) = rad(A) + Z(S)$. By theorem 10 a radical complement X of $rad(A)$ in S exists such that $nil(A°) \subseteq (X \cap Z(S))$ is valid. The set $X \cap Z(S)$ centralizes the nilradical of A. By a generalized version of the theorem of Wedderburn-Malcev (see e.g. [74]) X can be enhanced to a complement T of $rad(A)$ in A. Hence the set $X \cap Z(S)$ is contained in the centralizer of $rad(A)$ in T. Let $n \in nil(A°)$. Then there exist $r \in rad(A)$ and $v \in X \cap Z(S)$ such that $n = r + v$ is valid. The element v centralizes $rad(A)$. We have to prove that T is centralized by v, too. $rad(A)$ is contained in $nil(A°)$, and we derive $v = n - r \in nil(A°) \cap T$. This intersection is a nilpotent ideal of the Lie algebra $T°$, and T is as associative algebra semisimple. By our result for semisimple algebras this intersection is central in T (see theorem 15). The proof of the first part is finished as the sum of an ideal and a subalgebra is a subalgebra.

Add-on: For an arbitrary radical complement T of $rad(A)$ in A we have already proven that $nil(A°) \cap T$ is contained in $T \cap Z(A)$. Let $z \in Z(A)$. Then there exist $r \in rad(A)$ and $t \in T \cap Z(A)$ such that $z = r + t$ is valid. Hence $r = z - t \in Z(A) \cap rad(A)$ is true. Consequently, $Z(A)$ is the inner direct sum of $rad(A) \cap Z(A)$ and $T \cap Z(A)$. The first summand is a nilpotent ideal von $Z(A)$. The second summand is a central unital K-subalgebra of T. T is separable, and by results in [49] the center of T is a separable commutative K-algebra. By a theorem in [74] the subalgebra $T \cap Z(A)$ is separable, too (as above-mentioned). Thus $rad(A) \cap Z(A)$ is a nilpotent ideal with separable factor algebra by its nilradical of $Z(A)$ and must be identical to the radical of $Z(A)$. $Z(A)$ is commutative and separable, and the theorem of Wedderburn-Malcev implies that it possesses exactly one radical complement. This complement is exactly $T \cap Z(A)$. This argumentation is done for an arbitrary radical complement T and is valid for the intersection of all complements, too.◊

Example 2 We give an example such that theorem 16 is wrong in characteristic 2. Let $K := GF(2)$. We focus on the simple unitary finite-dimensional associative K-algebra $A := K^{2\times 2}$. It consists of 16 elements and has dimension 4. The center is exactly the set of all diagonal matrices with identical entries. A is a central algebra, and the center is of dimension 1. We prove that the nilradical of $A°$ three-dimensional and coincides with $A \circ A$. In addition, its a non-unital K-subalgebra of A.

Let a, b, c, d be the basis-matrices of $K^{2\times 2}$. (These are the matrices for which the only non-zero entry is 1.) Its straightforward to calculate $A \circ A = \langle b, c, a + d \rangle_K$. Every element in $A \circ A$ is ad-nilpotent on $A \circ A$, and thus – by a theorem of Engel – $A \circ A$ is nilpotent. $bc = a$ is valid, and thus $A \circ A$ is a non associative subalgebra (otherwise it would be A). In addition, $A°$

is not nilpotent proven by $a \circ b = b$. The reader may prove this example in details as an exercise.⋄

Remark 5 Let A be an associative K-algebra over a field K with $char(K) = p$. Then A° is – with respect to the powering with p – a so-called restricted Lie algebra. For this the nilradical is the greatest p-nilpotent ideal. By results in [65] (page 68, chapter 2, remark after 1.6) this nilradical is exactly the nilradical of A.⋄

4.6 Compatibility of the Lie nilradical with special constructions of algebras

In the previous sections we have determined the Lie nilradical for diverse classes of associative algebras: solvable, simple, semisimple, direct products etc. For direct products the determination was purely natural: its the direct product of the Lie nilradicals of each factor. In this section we analyze this question in more details: in what way is the determination of the Lie nilradical of special constructions of algebras to be answered naturally with respect to components of the construction?

4.6.1 Subalgebras

Let A be an associative unitary K-algebra and T an unital K-subalgebra. We focus on the identity $nil(T^\circ) = nil(A^\circ) \cap T$. The intersection is a nilpotent ideal of T° and contained in the Lie nilradical of T. The other inclusion is wrong in general: let $n \in \mathbb{N}$, K be a field with $char(K) \neq 2$, $A := K^{n \times n}$ and $T := \delta_{u,n}$. A is simple and therefor $nil(A^\circ) = Z(A) = K \cdot 1_A$ by theorem 16. T is a solvable associative K-algebra and its nilradical is the set of strict lower diagonal matrices. This set is contained in the nilradical of T° which cannot be identical to the center of A.⋄

4.6.2 Right and left ideals

Let A be an associative unitary K-algebra and R resp. L a K-right resp. K-left ideal of A. We analyze if for $T \in \{L, R\}$ the natural statement $nil(T^\circ) = nil(A^\circ) \cap T$ is true. The intersection is surely a nilpotent ideal of T° and hence contained in its Lie nilradical. But in general the complete identity is not true as the following example demonstrates: let $n \in \mathbb{N}$, K be a field with $char(K) \neq 2$, $A := K^{n \times n}$, L resp. R a one-line-space resp. one-column-space of A. (An one-line- resp. one-column-space is a set of matrices such that only in one line resp. one column an entry different from zero exists.) L and R are unital non-commutative K-subalgebras of A. Both are so-called local K-algebras and possesses an one-dimensional central radical complement. Therefor their associated Lie algebras are nilpotent. A is a simple algebra A° and by theorem 16 is its nilradical exactly its center.⋄

4.6.3 Ideals

Let A be an associative unitary K-algebra and I an ideal of A. Its natural to ask if $nil(I^\circ) = nil(I^\circ) \cap I$ is true. The intersection is clearly a nilpotent ideal of I° and thus contained in the Lie nilradical of I. But the whole identity is not correct commonly as the following example documents: we examine the Solomon-Tits algebra $A := K\Pi_n$ in $char(K) \neq 2$. Within proposition 8 in [76] an one-dimensional ideal I is constructed for which $AIA = I$ is proven. In particular, I is isomorphic to K and thus separable. Therefor – by using the Wedderburn-Malcev theorem – the ideal is contained in a radical complement of $rad(A)$ in A. If I is contained in $nil(A^\circ) \cap I = (rad(A) \oplus K \cdot 1) \cap I$, then a linear generator i of I could be presented as the sum of an element r of $rad(A)$ and $k \cdot 1_A$. The second summand is contained in every radical complement and thus r is nilpotent and separable. Consequently r is zero and $1_A \in I$. This is a contradiction because $I \neq A$ is valid (see theorem 16).⋄

4.6.4 Factor algebras

Let A be an associative unitary K-algebra, $char(K) \neq 2$ and I an ideal of A. We analyze whether the statement $nil((A/I)^\circ) = (nil(A^\circ) + I)/I$ is true. The right side of the equation is surely a Lie nilpotent and is therefor contained in $nil((A/I)^\circ)$. The following example shows us that the identity is not true in general: we focus again on the Solomon-Tits algebra $A := K\Pi_n$. By results in [76] we know that A is solvable which means that $A/rad(A)$ is commutative. $A/rad(A)$ is Lie nilpotent. But $nil(A^\circ) = rad(A) \oplus K \cdot 1_A$ is valid and we conclude that $(nil(A^\circ) + rad(A))/rad(A)$ is of dimension 1 and not identical to $A/rad(A)$ (see theorem 16).⋄

4.6.5 The opposite algebra

Let A be an associative algebra. The opposite algebra A^{op} or inverse algebra A^- is defined by the multiplication $a \cdot_{op} b := ba$ for all $a, b \in A$. It is straightforward to prove that the multiplication of the associated Lie algebras A° and $(A^{op})^\circ$ differs only by $-$. Hence we conclude $nil(A^\circ) = nil((A^{op})^\circ)$.⋄

4.6.6 Matrix algebras

Let A be an associative unitary K-algebra, $n \in \mathbb{N}$ and I_n the unit of $A^{n \times n}$. We analyze whether the statement $nil((A^{n \times n})^\circ) = (nil(A^\circ))^{n \times n}$ is valid. We determine the Lie nilradical of matrix algebras, prove one inclusion of our statement and show that the identity is wrong in general.

Proposition 7 Let K be a field, $n \in \mathbb{N}$, A an associative finite-dimensional unitary K-algebra with separable factor algebra by the nilradical and T a radical complement of $rad(A)$ in A. The subalgebra $T^{n \times n}$ is a separable radical complement of $rad(A^{n \times n}) = rad(A)^{n \times n}$ in $A^{n \times n}$. In particular, $A^{n \times n}$ is an associative finite-dimensional unitary K-algebra with separable factor algebra by the nilradical.

Proof. It is well-known and often proven by using Morita[4]-theory (see e.g. [49]) that $rad(A^{n \times n}) = rad(A)^{n \times n}$ is valid. $T^{n \times n}$ is a complement for this nilradical. We have to prove that this complement is separable. For this it is equivalent to show that $X := T^{n \times n} \otimes (T^{n \times n})^{op}$ is semisimple (see e.g. [49]). We use several exercises presented in section 6.2 to determine the tensor product X. $T^{n \times n}$ is isomorphic to $T \otimes K^{n \times n}$, and $(T^{n \times n})^{op}$ is isomorphic to $(T^{op})^{n \times n}$. Hence X is isomorphic to $(T \otimes T^{op}) \otimes K^{n \times n} \otimes K^{n \times n}$. The last two factors are isomorphic to $K^{n^2 \times n^2}$. Therefor X is isomorphic to $(T \otimes T^{op})^{n^2 \times n^2}$. T is separable and we derive that $T \otimes T^{op}$ is semisimple. By using the result about the nilradical (as above-mentioned) of matrix algebras again we conclude that X is semisimple. $\diamond$

Theorem 17 Let K be a field with $char(K) \neq 2$, $n \in \mathbb{N}$, A an associative finite-dimensional unitary K-algebra with separable factor algebra by the nilradical and T a radical complement of $rad(A)$ in A. The following identities are valid:

(i) $nil((A^{n \times n})^\circ) = rad(A)^{n \times n} \oplus (Z(A) \cap T) \cdot I_n$

(ii) $nil((A^{n \times n})^\circ) \subseteq (nil(A^\circ))^{n \times n}$

(iii) $nil((A^{n \times n})^\circ) = (nil(A^\circ))^{n \times n}$ is wrong for all $n \neq 1$.

Proof. It is well-known that $Z(A) \cdot I_n$ is the center of $A^{n \times n}$. Therefor the parts (i) and (ii) are a consequence of theorem 16 and of proposition 7. Part (iii) is valid as the identity is only true if $K^{n \times n}$ is Lie nilpotent. This is not true for $char(K) \neq 2$ which the reader shall prove as an exercise. $\diamond$

4.6.7 Adjunction of a unit

In this short section we enhance theorem 16 to non-unitary K-algebras. Within this we derive that the Lie nilradical of the adjunction of a unit can be described by the underlying algebra. In definition 3 the algebra A^K is the adjunction of a unit.

---

[4] Kiiti Morita, born February 11, 1915, died August 4, 1995 was a Japanese mathematician working in algebra and topology. Born in Hamamatsu, he received his Ph.D. from the University of Osaka in 1950 and was professor at the University of Tsukuba. He introduced the concepts now known as Morita equivalence and Morita duality which were given wide circulation in the 1960s by Hyman Bass in a series of lectures. The Morita conjectures on normal topological spaces are also named after him.

Theorem 18 *Let A be a finite-dimensional associative K-algebra with separable factor algebra by its nilradical, T a radical complement and K a field with $char(K) \neq 2$. Then $nil(A^\circ)$ is an associative subalgebra of A and the identities $nil(A^\circ) = rad(A) \oplus (Z(A) \cap T)$ and $nil(A^\circ)^K = nil((A^K)^\circ)$ are valid.*

Proof. By using results in [74] we conclude that A^K is satisfying the preconditions of theorem 16, T^K is a radical complement and $rad(A) = rad(A^K)$ is valid. By proposition 13 the algebra $Z(A)^K$ is the center of A^K, and its straightforward to prove that $(nil(A)^\circ)^K$ is a nilpotent ideal of $(A^K)^\circ$. Theorem 16 implies that $rad(A) \oplus (T \cap Z(A))^K$ is the Lie nilradical of A^K. ⋄

4.6.8 Tensor products

In this section we analyze the determination of the Lie nilradical of a tensor product of associative algebras. In particular, we focus on the connection of this nilradical to the Lie nilradical of each factor. Let A, B associative finite-dimensional unitary K-algebras with separable factor algebras by their nilradical and T_A resp. T_B a radical complement of $rad(A)$ resp. $rad(B)$ in A resp. B. We assume $char(K) \neq 2$. By a result in [74] the subalgebra $T_A \otimes T_B$ is a radical complement of $rad(A \otimes B)$ in $A \otimes B$. In addition, the identities $rad(A \otimes B) = rad(A) \otimes B + A \otimes rad(B)$ and $Z(A \otimes B) = Z(A) \otimes Z(B)$ are valid. By using theorem 16 we derive the following results:

(i) $nil(A^\circ) = rad(A) \oplus (Z(A) \cap T_A)$

(ii) $nil(B^\circ) = rad(B) \oplus (Z(B) \cap T_B)$

(iii) $nil((A \otimes B)^\circ) = (rad(A) \otimes B + A \otimes rad(B)) \oplus ((Z(A) \otimes Z(B)) \cap (T_A \otimes T_B))$

(iv) $nil(A^\circ) \otimes nil(B^\circ) = (rad(A) \oplus (Z(A) \cap T_A)) \otimes (rad(B) \oplus (Z(B) \cap T_B))$.

By using parts (iii) and (iv) we conclude that $nil(A^\circ) \otimes nil(B^\circ)$ is contained in $nil((A \otimes B)^\circ)$. Again by (iii) and (iv) and by $A = T_A \oplus rad(A)$ resp. $B = T_B \oplus rad(B)$ the statements (iii) and (iv) are identically if and only if T_A and T_B are central. This statement is equivalent (by results in [74]) that A° and B° or that $(A \otimes B)^\circ$ is nilpotent. ⋄

4.7 The group algebra once again

We have determined the Lie nilradical for the Solomon-Tits algebras, for the Solomon algebras in zero characteristic and the algebras of upper and lower triangular matrices over a field. In addition, we have focussed on solvable group algebras and determined the Lie nilradical, too, but with a complete different argumentation as used for the other algebras which have a self-centralizing radical complement. We prove that group algebras do not belong to this kind of algebras and use theorem 16 to derive some general results for its Lie nilradical.

Remark 6 Let K be a field and G a finite group.

Let KG be semisimple. Then the nilradical of KG is the zero space and KG is the only radical complement. KG is self-centralizing if and only if KG is commutative and G is abelian.

The more interesting case is if KG is not semisimple. Hence $char(K) = p$ is a prime number dividing $|G|$. Using results of Karpilovsky in [33] and in [34] we apply that $KG/rad(KG)$ is separable. Therefor $rad(KG)$ possesses (theorem of Wedderburn-Malcev) a radical complement T. We assume that T is self-centralizing. Then we would derive $Z(KG) \leq T = C_{KG}(T)$. T is separable and we conclude by using standard results in [49] that the center of T is a separable commutative K-algebra. This would imply by using results in [74] that the center is – as a subalgebra of T – separable and commutative, too. The sum of all elements of G is a nilpotent and central element of KG. This element is nilpotent and separable and this implies that its the zero element which is a contradiction.◇

We have used a result of Karpilovsky (in [33] and in [34]) that the factor algebra of the nilradical in KG is separable for every finite group and arbitrary field. By using theorem 16 we get the following theorem:

Theorem 19 Let K be a field with $char(K) \neq 2$ and G a finite group. Then the unital subalgebra $rad(KG) + Z(KG)$ of KG is exactly the nilradical of $(KG)^{\circ}$.

Add-on: $rad(Z(KG)) = rad(KG) \cap Z(KG)$, $Z(KG)/rad(Z(KG))$ is separable and for every radical complement T of $rad(KG)$ in KG the subalgebra $T \cap Z(KG) = (\bigcap_{r \in rad(KG)} T^{1+r}) \cap Z(KG)$ is the radical complement of $rad(Z(KG))$ in $Z(KG)$. In particular, for every radical complement T of $rad(KG)$ in KG the identity $nil((KG)^{\circ}) = rad(KG) \oplus (Z(KG) \cap T)$ is valid. ◇

Sections 4.5 - 4.7

$\text{nic}(A^9) = g(A) \oplus \mathbb{Z}$
$g(A)$
$g(\mathbb{Z}(A))$
$\mathbb{Z}(A) \cap g(A)$
$g(A)$
$\mathbb{Z}(A)$
$G(T)$?
$T \oplus g(\mathbb{Z}(A))$
$\mathbb{Z} = \mathbb{Z}(A) \cap T$

T separate

$g_{n \times n}$
$\text{nic}(A_{n \times n}^9)$
$(\mathbb{Z}(A) \cap T)$
$T_{n \times n}$
$A_{n \times n}$

Theorem 4.5.1
- Not compliant with n, θ and /
- Compliant with θ:
  $\text{nic}(A^9 \theta) = \text{nic}(A^9) \theta$
- Compliant with x:
  $\text{nic}(A \times S)^9 = \text{nic}(A^9) \times \text{nic}(S^9)$
- Compliant with 4:
  $\text{nic}(A^4)^9 = \text{nic}(A^9)^4$
- Compliant with the $n \times n$:
  $\text{nic}((A^{n \times n})^9) = g(A)^{n \times n} \oplus (\mathbb{Z}(A) \cap T) \cdot \mathbb{1}_{n \times n}$
- Compliant with group algebra:
  T separate, always exists!

4.8 Open-ended questions and exercises

Open-ended questions 1 *(i) How can we determine the nilradical of an associated Lie algebra of an arbitrary associative algebra?*

(ii) How can we determine the solvable radical of an associated Lie algebra of an arbitrary associative algebra?

(iii) What are the answers of theses two questions for the associated Jordan algebra of an arbitrary associative algebra?

(iv) What are the answers for (i) and (ii) related to the p-structure of the restricted Lie algebra?

(v) What are the maximal p-nilpotent Lie subalgebras in the case $char(K) = p$?

(vi) Is the intersection of all radical complements of an arbitrary associative algebra central (This was proven for the solvable case.) or has it a different structural meaning?

(vii) What is the nilpotency and solvable class of the nilradical of an associated Lie algebra based on an associative unitary finite-dimensional algebra?

Excercise 37 *Let p be a prime number, $n \in \mathbb{N}$, K a field and $G \in \{D_{2p^n}, D_{4p^n}, Q_{4p^n}\}$. Determine the dimension, nilpotency and solvable class of the nilradical of $(KG)^\circ$. If this exercise is to complex, then let $p \in \{2,3,5,7\}$!*

Excercise 38 *Let K be a field and A an associative finite-dimensional solvable unitary K-algebra with separable factor algebra $A/rad(A)$. One self-centralizing radical complement does exist. Prove that all radical complements are self-centralizing and maximal commutative subalgebras of A. In addition, analyze whether the algebras of lower and upper triangular matrices over a field, the Solomon algebra in characteristic zero or the Solomon-Tits algebras are examples for this result. (Tip: [76])*

Excercise 39 *(zero-extension) Let A be a K-algebra with respect to the multiplication $\cdot$. We define a multiplication $\odot$ on $B := A \times A$ by $(a,x) \odot (b,y) := (ab, ay+xb)$. Analyze the following topics:*

(i) $(B; \odot)$ is a K-algebra.

(ii) $(A; \cdot)$ is associative if and only if $(B; \odot)$ is associative.

(iii) $(A; \cdot)$ is commutative if and only if $(B; \odot)$ commutative.

(iv) If $(A; \cdot)$ is unitary, then $(B; \odot)$ is unitary, too, and $(1_A; 0)$ is its identity element.

(v) Is the opposite implication of (iv) true?

(vi) The center of $(B; \odot)$ is $Z(A) \times Z(A)$.

(vii) $0 \times A$ is a nilpotent ideal of B with trivial multiplication (a so-called zero-ideal).

(viii) $B/(0 \times A)$ is isomorphic to A.

(ix) Let A be associative unitary finite-dimensional and separable. Prove that B is a unital finite-dimensional associative K-algebra with nilradical $0 \times A$. The factor algebra by the nilradical is isomorphic to A. $A \times 0$ is a radical complement. Is it isomorphic to A, too?

(x) Use theorem 16 and the results of this exercise to determine the nilradical of B° in the case $char(K) \neq 2$ in the previous part.

(xi) Is B isomorphic to the direct product $A \times A$?

(xii) Drop down the separability of A in part (ix). Is $(rad(A); A)$ the nilradical of B? Is the factor algebra by the nilradical isomorphic to $A/rad(A)$? Generalize the determination of the nilradical of B° by using this and theorem 16!

Excercise 40 Let K be a field and $n \in \mathbb{N}$. Analyze whether $gl(n, K)$ is nilpotent or solvable.

Excercise 41 Prove example 2 in details!

Excercise 42 Based on remark 3 summarize the results and formulate a theorem!

Excercise 43 Use the results of remark 3 for the following cases:

(i) G is a p-group and $char(K) = p$ is valid.

(ii) G is a dihedral group and $char(K)$ is fitting.

(iii) G is a quaternion group and $char(K)$ is fitting

(iv) G is a semi dihedral group and $char(K) = 2$ is valid.

Excercise 44 Two normal subgroups with trivial intersection centralize each other!

Excercise 45 Within remark 3 the identity $Z(KG) \cap KH = Z(KG) \cap (KH)^x$ for a unit x of KG is used. Prove this identity. In addition, prove $core_G(H) = core_G(H^g)$ for all $g \in G$.

Excercise 46 Let G be a group and H a subgroup of G. Prove that the core of H in G (which is the greatest normal subgroup of G in H) is exactly the intersection of all conjugates of H in G. Find a group homomorphism for which the kernel is exactly the core of H in G. (Tip: Operation on the cosets of H in G!)

Excercise 47 Let A be an associative unitary K-algebra, r a nilpotent element of A, T, D K-subalgebras and I a K-ideal of A. Prove that the identities $C_T(D)^{1+r} = C_{T^{1+r}}(D^{1+r})$ and $I^{1+r} = I$ are valid. What is the corresponding identity in a non-unital algebra?

Excercise 48 Let A be a 3-dimensional K-algebra with Basis $\{1, z, w\}$ and the following multiplication rules: 1 is neutral, $z^2 = z$, $zw = w$, $wz = ww = 0$. Prove that A is associative and determine the nilradical of A°.

Excercise 49 Let A be a unitary associative algebra for which every element is idempotent. Determine the nilradical of its associated Lie algebra (Tip: square $2 * 1_A$ and $a + b$). Do you know prominent examples for this kind of algebras? In what way are central idempotents an example? Find the name for such algebras. What do we know about the mathematician these algebras are named of?

Excercise 50 Formulate and prove the following theorem (see also theorem 16): The add on is a consequence of the statement that a sum of an ideal and a subalgebra is a subalgebra, too.

Excercise 51 Let A be an associative K-algebra and I a K-ideal of A. We define $S_I := \{s \mid s \in A, \forall a \in A : s \circ a \in I\}$. Analyze whether S_I is a K-subalgebra or a K-ideal of A. Is S_I a K-subalgebra or K-ideal of A°? What is the importance of S_I/I in A/I or in A°/I? What is the purpose of S_I for theorem 16? What are the results if I is only an ideal of A°?

Excercise 52 Let A be an associative K-algebra, I an ideal of A and T be a K-subalgebra of A. Prove that $I \cap T$ is an ideal of T°. If I is nilpotent, then $I \cap T$ is nilpotent, too.

Excercise 53 Let K be a field and A a finite-dimensional associative unitary K-algebra with separable factor algebra by the nilradical. Is the nilradical of A° a (K)-ideal, a left or right (K)-ideal or an (unital) (K)-subalgebra of A?

Excercise 54 Prove proposition 15 in details!

Excercise 55 Which statements of proposition 15 are still valid if the word K-ideal is replaced by the word left K-ideal, right K-ideal or (unital) K-subalgebra?

Excercise 56 *Prove remark 4 in details!*

Excercise 57 *Let K be a field, A an associative unitary K-algebra and e an idempotent of A. Determine a generalized Jordan decomposition of $ad(e)$!*

Excercise 58 *Let A be an associative K-algebra and $(r;s)$ a generalized Jordan decomposition. Prove that $(r\rho; s\rho)$ and $(r\lambda; s\lambda)$ are generalized Jordan decompositions (of which elements?).*

Excercise 59 *Let A be an associative K-algebra, $(r;s)$, (u,v) generalized Jordan decompositions and $k \in K$. On what terms are $(kr; ks)$, $(r+u; s+v)$ and $(ru; sv)$ generalized Jordan decompositions (of which elements?)?*

Excercise 60 *Let A be an associative unitary K-algebra and $a \in A$. Prove that the minimal polynomial of a, $\lambda(a)$ and $\rho(a)$ are identically. In addition, prove that for all $r, s \in A$ the identity $ad(r + s) = ad(r) + ad(s)$ is valid. Calculate $ad(r \circ s)$! On what terms on r and s possesses this element a generalized Jordan decomposition?*

Excercise 61 *Let A be a K-algebra, I an ideal and T a subalgebra of A. Prove that $I + T$ is a subalgebra of A.*

Excercise 62 *Let K be a field, $n \in \mathbb{N}$ and $A := K^{n \times n}$ the algebra of $n \times n$-matrices over K. Determine the center of A! Determine the nilradical of A and of A°!*

Excercise 63 *Let A be a unitary associative solvable finite-dimensional K-algebra and N the nilradical of A°. Determine the nilradical of A°/N? Is the result also valid for an arbitrary Lie algebra?*

Excercise 64 *Let A be an associative solvable K-algebra. Every subalgebra of A° containing $rad(A)$ is an ideal of A°.*

Excercise 65 *Let A be an associative K-algebra and L a maximal abelian Lie subalgebra of A°. Prove that L is an associative and commutative subalgebra of A. (Tip: For all $a \in A$ the function $ad(a)$ is a derivation. If $a, b \in L$, then focus on the K-span of $L \cup \{ab\}$ which is an abelian Lie subalgebra of A° containing L.) What is the consequence for the maximal commutative subalgebras of A and the maximal abelian subalgebras of A°?*

Excercise 66 *Prove the previous exercise again by usage of the following idea: If A is an associative K-algebra and T a set of pairwise commutable elements, then the algebra-span of T is a commutative subalgebra of A.*

Excercise 67 *Let K be a field and $A := K[t]$ the polynomial algebra over K. For the following polynomials determine the greatest common divisor gcd, the lowest common multiple lcm and their formal derivation:*

(i) $K = \mathbb{R}$, $f = t^2 + t + 1$, $g = t - 1$

(ii) $K = \mathbb{Q}$, $f = t^3 + t$, $g = t(t-1)$

(iii) $K = \mathbb{C}$, $f = t^4 + t^2 + 1$, $g = t^2 + 1$

(iv) $K = GF(p)$, $f = t^p + t + 1$, $g = t - 1$

(v) $K = GF(p)$, $f = t^{p^2} + t^p + 1$, $g = t$.

Excercise 68 Let K be a field and $A := K[t]$ the polynomial algebra over K. Are the following polynomials nilpotent, semisimple, fully separable, separable, irreducible?

(i) $K = \mathbb{R}$, $f = t^2 + t + 1$, $g = t - 1$

(ii) $K = \mathbb{Q}$, $f = t^3 + t$, $g = t(t-1)$

(iii) $K = \mathbb{C}$, $f = t^4 + t^2 + 1$, $g = t^2 + 1$

(iv) $K = GF(p)$, $f = t^p + t + 1$, $g = t - 1$

(v) $K = GF(p)$, $f = t^{p^2} + t^p + 1$, $g = t$.

Excercise 69 Let K be a field and $A := K[t]$ the polynomial algebra over K. Is the formal derivation of polynomials from A in A surjective, injective, bijective, K-linear or multiplicative? Is there a special result for finite or algebraically closed K?

Excercise 70 Prove the preliminary remark 1 in details!

Excercise 71 Let K be a field and A an associative finite-dimensional solvable K-algebra with separable factor algebra by its nilradical. If T is a subalgebra of A, then T is an associative finite-dimensional solvable K-algebra with separable factor algebra by its own nilradical. (Tip: [74], chapter 3+5).

Excercise 72 Let p be a prime number, K a field of characteristic p and G a finite non-abelian group such that KG is solvable. Prove that G possesses a normal p-Sylow subgroup, and the class of nilpotency of $rad(KG)$ and $Aug(KP)$ is identical. Determine this nilpotency class for a group of order p^3.

Excercise 73 Define the term Lie algebra and prove in details that the associated Lie algebra of an associative algebra is a Lie algebra!

Excercise 74 What would be the consequence if we define the adjoint representation of a Lie algebra not by 'xl' but by 'lx'?

Excercise 75 *We focus on the Lie algebra $gl(2,p)$. Let a,b,c,d the basis matrices of $K^{2\times 2}$ and $B := \{a,b,c,d\}$. What is the set $B \circ B$? What is the dimension of $\langle B \circ B \rangle_{GF(p)}$? Is the identity $\langle B \circ B \rangle_{GF(p)} = \langle B \circ B \rangle_{\mathbb{Z}}$ valid? Calculate the minimal polynomial and the characteristic polynomial of all elements of B! What is the generalized Jordan decomposition of the elements of B? On what terms is $ad(x)$ (for $x \in B$) nilpotent or fully-separable? Calculate $ad(x)^p$ for all $x \in B$! Represent this linear function by the basis B and determine its generalized Jordan decomposition! Do we have any special results for the case $p = 2$?*

Excercise 76 *Within remark 3 prove that for an element $p \in P$ the conditions $C_G(p)$ is normal, P is contained in $C_G(p)$ and p does not divide $\mid p^G \mid$ are equivalent. Find a sufficient conditions for P such that one of these three statements is valid!*

Chapter 5

Cartan subalgebras in Lie algebras associated to associative algebras

In the previous chapter we have deeply investigated the nilradical of Lie algebras associated to associative algebras and described its structure as an associative subalgebra. The nilradical is one special member of the set all maximal Lie nilpotent substructures. Other important ones are the Cartan subalgebras which are in focus within this chapter. They are defined as nilpotent and self-normalizing subalgebras of a Lie algebra. In our context Cartan subalgebras do exist. We will show that they are associative subalgebras and analyze the associative structure more deeply. Some of the results of this chapter are based on an article by Salvatore Siciliano [59], others are enhancements of his theory which we will widen to other classes of algebras (like basic algebras or algebras with separable factor algebra by its nilradical). Finally, we will analyze our standard examples, the group algebras, the Solomon-Tits algebras, the Solomon algebras in characteristic zero and the algebras of upper and lower triangular matrices over an arbitrary field, concerning Cartan subalgebras.

5.1 Cartan subalgebras are associative solvable subalgebras

In this section we focus on the associative structure of Cartan subalgebras in Lie algebras associated to associative algebras. We will prove that these Lie subalgebras are indeed associative subalgebras and, in addition, that they are solvable. For a perfect field we prove further that every fully separable element is central. Therefor exactly one radical complement exists which is central. The main result for nilradicals in chapter 4 play an important role for proving this result.

5.1.1 Associative structure of Cartan subalgebras

Salvatore Siciliano proved in [59], theorem 1 in particulary that for a finite-dimensional associative unitary K-algebra every Cartan subalgebra of its associated Lie algebra is indeed an associative subalgebra. This result is true in a more general context.

Definition 1 (i) For all $n \in \mathbb{N}$ let $\underline{n} := \mathbb{N}_{\leq n}$.

(ii) For all $n \in \mathbb{N}_{\geq 2}$ we define
$T_n := \{(\alpha, \beta) \mid \exists r \in \underline{n-1}, a_1, ..., a_n \in \underline{n} : \alpha = (a_1, ..., a_r),$
$\beta = (a_{r+1}, ..., a_n), a_1 < ... < a_r, a_{r+1} < ... < a_n, \underline{n} = \{a_1, ..., a_n\}\}.$

(iii) For a nilpotent Lie algebra L let $cl(L)$ be its class of nilpotency.

(iv) For an associative K-algebra A and subsets T, S of A let $C_A(T)$ be the centralizer of T in A and $C_S(T) := C_A(T) \cap S$.

(v) Let A be an associative K-algebra. For a subset T of A let $\langle T \rangle_K$ resp. $\langle T \rangle_A$ resp. $\langle T \rangle_{A_1}$ be the K-linear resp. algebra resp. unital algebra span of T in A.

(vi) If L is a Lie algebra and T a subset of L, then let $N_L(T)$ be the normalizer of T in L.⋄

Remark 7 Let A be an associative K-algebra and $a, b \in A$.

(i) For all $h_1 \in A$ the following derivation rule is valid:
$$(ab)\,ad(h_1) = (a\,ad(h_1))b + a(b\,ad(h_1)).$$

(ii) For all $n \in \mathbb{N}_{\geq 2}, h_1, ..., h_n \in A$ we obtain by (i) and induction:
$$(ab)ad(h_1)...ad(h_n)$$
$$= (a\,ad(h_1)...ad(h_n))b + a(b\,ad(h_1)...ad(h_n))$$
$$+ \sum_{((\alpha_1,...,\alpha_r),(\alpha_{r+1},...,\alpha_n)) \in T_n} (a\,ad(h_{\alpha_1})...ad(h_{\alpha_r}))(b\,ad(h_{\alpha_{r+1}})...ad(h_{\alpha_n})).\diamond$$

Proposition 8 *Let A be an associative (unitary) K-algebra and C a Cartan subalgebra of A°. Then C is a (unital) associative subalgebra of A.*

Proof. As Lie subalgebra of A° the algebra C is a K-space of A. Let $a, b \in C$. Then we have to prove that $ab \in C = N_{A^\circ}(C)$ is valid. This condition is equivalent to $(ab)ad(h_1) \in C = N_{A^\circ}(C)$ for all $h_1 \in C$. By induction we have to prove that an element $n \in \mathbb{N}$ has to exist such that for all $h_1, ..., h_n \in C$ the identity $(ab)ad(h_1)...ad(h_n) \in C$ is valid. We define $n := 2 \cdot cl(C)$. Part (ii) of remark 7 implies that $(ab)ad(h_1)...ad(h_n) = 0 \in C$ (One factor of summands within the sum is always equal to zero.) is valid. If A is unitary, then $1_A \in N_{A^\circ}(C) = C$ is valid.⋄

Remark 8 Let A, B be associative K-algebras, H a subalgebra of A and L a superfield of K.

(i) For all $a_1, a_2 \in A, b_1, b_2 \in B$ the identity

$$(a_1 \otimes b_1) \circ (a_2 \otimes b_2) = (a_1 \circ a_2) \otimes (b_1 b_2) + (a_2 a_1) \otimes (b_1 \circ b_2)$$

is valid.

(ii) If A° is nilpotent and B° is abelian, then we deduce by induction and part (i) that $(A \otimes B)^\circ$ is nilpotent.

(iii) The dimension, nilpotency and solvable class of H as K-algebra and $H \otimes L$ as L-algebra are identical.⋄

Remark 9 Let K be a field and $n \in \mathbb{N}$. We prove that $gl(n, K) := (K^{n \times n})^\circ$ is not nilpotent for $n \geq 2$. $gl(n, K)$ contains for $n \geq 2$ a subalgebra isomorphic to $gl(2, K)$. Hence we need only to show that $gl(2, K)$ is not nilpotent. Let

$$e_{11} := \begin{pmatrix} 1 & 0 \\ 0 & 0 \end{pmatrix} \text{ and } e_{12} := \begin{pmatrix} 0 & 1 \\ 0 & 0 \end{pmatrix}.$$

Then $e_{11} e_{12} = e_{12}$ is valid, and for all $r \in \mathbb{N}$ we deduce $e_{12} ad(e_{11})^r = (-1)^r e_{12} \neq 0$.⋄

Lemma 2 *Every finite-dimensional associative Lie nilpotent K-algebra is solvable.*

Proof. Step 1: Let A be a central K-division algebra, $n := ind(D)$ and T a maximal subfield of A. It is well-known that $A \otimes T$ and $T^{n \times n}$ are isomorphic. By using remarks 8 and 9 we deduce $A = K1_A$.

Step 2: Let A be a K-division algebra. A is as $Z(A)$-algebra central and again Lie nilpotent. Lie nilpotency is independent on the base field. Step 2 follows from step 1, and we conclude that $A = Z(A)$ is valid.

Step 3: Let A be simple. Then there exists a K-division algebra D and a natural number $n \in \mathbb{N}$ such that A is isomorphic to $D^{n \times n}$. A is Lie nilpotent, thus D has the same property and by step 2 the division algebra D is a field. In addition, A° contains a subalgebra isomorphic to $gl(n, K)$. By using remark 9 we deduce $n = 1$, and hence A is a field.

Step 4: Let A be semisimple and thus a direct product of simple ideals of A. By using step 3 all these ideals are fields and A is commutative.

Step 5: By using $A^\circ/rad(A)^\circ = (A/rad(A))^\circ$ the algebra A is solvable applying the result of step 4 to this factor algebra. ◇

Lemma 2 and proposition 8 imply:

Corollary 8 *Let A be a finite-dimensional associative (unitary) K-algebra. Every Cartan subalgebra of A° is a solvable associative (unital) subalgebra of A.* ◇

In the next corollary we obtain more information about the associative structure of Cartan subalgebras. Within the proof we use the main result about nilradicals of Lie algebras from chapter 4.

Corollary 9 *Let A be a finite-dimensional associative unitary K-algebra, K a perfect field and C a Cartan subalgebra of A°. The set of fully separable elements of C is central in C and the unique radical complement of $rad(C)$ in C.*

Proof. By using corollary 8 the Lie algebra C is an associative solvable unitary K-algebra, and by the perfectness of K the factor algebra $C/rad(C)$ is separable. Theorem 10 and the Lie nilpotency of C imply that C possesses a radical complement of elements in C which are central and fully separable in C ($A = C_A(T)$ is equivalent to $T \leq Z(A)$). By using the theorem of Wedderburn-Malcev the algebra C possesses exactly one radical complement (because all are conjugated and one is central). Let a be a fully separable element of C. Results in [74] let us deduce that the algebra $\langle a \rangle_{A_1}$ is separable, and thus – by an enhancement of the theorem of Wedderburn-Malcev – included in a radical complement of C. The proof is finished as this complement is unique. ◇

5.1.2 Open-ended questions and exercises

Open-ended questions 2 *(i) For every associative algebra over a field of characteristic $\neq 2$ by the multiplication $a \circ_1 b := ab + ba$ for all $a, b \in A$ a Jordan algebra A°_1} is defined – the Jordan algebra associated to A. Is it possible to transfer proposition 8 to A°_1}?*

(ii) Is it possible to transfer the results of this work for associated Lie algebras to associated Jordan algebras?

Excercise 77 *Prove remark 7.*

Excercise 78 For all $n \in \underline{6}$ determine the set T_n.

Excercise 79 Let A be an associative K-algebra. A derivation of A is a K-linear function d on A such that for all $a, b \in A$ the condition $(ab)d = (ad)b + a(bd)$ is valid. Prove that the set of all derivations of A with respect to $\circ$ is a Lie algebra, too. The inner derivations $\mathrm{ad}(a), a \in A$ form a subalgebra of this Lie algebra. Is this subalgebra an ideal?

Excercise 80 Let K be a field and $n \in \mathbb{N}$. The Lie algebra associated to $K^{n \times n}$ is denoted by $gl(n, K)$. Prove that the set of diagonal matrices is a Cartan subalgebra of $gl(n, K)$. What is class of nilpotency of this Cartan subalgebra? What is the dimension of $gl(n, K)$ and of this Cartan subalgebra?

Excercise 81 Let K be a field and $n \in \mathbb{N}$. Is the set of diagonal matrices a Cartan subalgebra of the Lie algebra of lower triangular matrices?

Excercise 82 Let K be a field and $n \in \mathbb{N}$. What is the class of nilpotency of the strict lower triangular matrices as subalgebra of $gl(n, K)$? What is the normalizer of this subalgebra in $gl(n, K)$ and in the Lie algebra of lower triangular matrices?

Excercise 83 Formulate the following statement and prove it: If a Cartan subalgebra is contained in a subalgebra T of the underlying Lie algebra L, then the Cartan subalgebra T is a Cartan subalgebra of T.

Excercise 84 Let K be a field and $n \in \mathbb{N}$. The set of matrices with trace equal to zero is a subalgebra of $gl(n, K)$. Is the set of diagonal matrices with trace equal to zero a Cartan subalgebra of this subalgebra?

Excercise 85 Let K be a field. What is the center of $K^{3 \times 3}$? Determine a non-central matrix A of $gl(3, K)$ and calculate the minimal and characteristic polynomial of $\mathrm{ad}(A)$!

Excercise 86 Let K be a field. What is the center of $K^{4 \times 4}$? Determine non-central pairwise distinct matrices A, B, C of $gl(4, K)$ and calculate $(AB)\mathrm{ad}(C)^r$ for all $r \in \underline{4}$!

Excercise 87 Let K be a field and D a finite-dimensional associative K-division algebra. Prove that $Z(D)$ is a field and D is a central $Z(D)$-algebra. In addition, D° is as K-algebra nilpotent resp. solvable if and only if it has these properties as a $Z(D)$-Algebra.

Excercise 88 Let A be an associative K-algebra and I an ideal of A. Prove that I° is an ideal of A° and $A^\circ / I^\circ = (A/I)^\circ$ is valid.

Section 1.1.

> associative structure of
> Coatan subalgebra

C Coatan subalgebra   C is unital and subalgebra
of A   associative subalgebra

            ⟶ { for a
                perfect field

C Coatan subalgebra ~ = mult(C)
of A   = $g(C) \otimes T(C)$

                        A°
                       •————————————•
            $g(C)$ /   /
                     •————————————•
                    a $T(C)$

set of $g(C)$ = set of all jointly separable elements
all
nilpotent   ~ unique radical
elements of C   complement of C
of C
                       ~ VSEPR(C) in SL2

5.2 Maximal tori and Cartan subalgebras

This section is based on the article [59] of Salvatore Siciliano. I want to say thank you to Salvatore Siciliano for his helpful comments for creating this section. We describe how Cartan subalgebras in Lie algebras associated to associative algebras can be constructed by maximal tori. Both structures – Cartan subalgebras and maximal tori – are closely connected. This relation is one of the main results of this work.

We demonstrate the main result by some of our standard examples, the Solomon algebras, the Solomon-Tits algebras, the algebra of lower and upper triangular matrices over a field and some special solvable group algebras based on dihedral and quaternion groups.

5.2.1 Maximal tori

Definitions and remarks 3 Let A be a K-algebra over a field K of positive characteristic p and S a subset of A. By $\langle S \rangle_p$ we denote the restricted Lie subalgebra of A° generated by S. A° is with respect to the p-function $x \mapsto x^p$ restricted. If we focus on A as a restricted Lie algebra, then we denote it by $(A^\circ; [p])$ (see e.g. [65] for a detailed description of restricted Lie algebras). If M is a set, then id_M is the identity function on M.⋄

Fully separable elements are defined for restricted Lie algebras, too (see e.g. [65]). The next results shows that our description is compatible with that definition:

Proposition 9 *(Siciliano) Let A be an associative K-algebra over a field K of positive characteristic p. An element x of A is fully separable if and only $x \in \langle x^p \rangle_p$ is valid.*

Proof. Let F be an algebraically closed superfield of K and x be fully separable. By using definition and remark 3 the element $x\rho$ is fully separable, too. Hence $x\rho \otimes id_F$ is a diagonizable endomorphism of $A \otimes_K F$. Thus $\langle x\rho \otimes id_F \rangle_A$ contains no non-zero nilpotent element. By [81] (or [65], section 2.3) we deduce that $x\rho \in \langle x\rho^p \rangle_p$ is valid. $x = 1(x\rho)$ implies $x \in \langle x^p \rangle_p$.
Now let $x \in \langle x^p \rangle_p$. Then elements $a_1, a_2, \ldots, a_r \in K$ exist such that $x = \sum_{i=1}^{r} a_i x^{p^i}$ is valid. We define the polynomial $f = t - \sum_{i=1}^{r} a_i t^{p^i}$. It is straightforward to prove that $f(a) = 0$ is valid, and this implies $min_{x,K} \mid f$. The formal derivation of f is – by using $char(K) = p$ – exactly $f' = 1$. Finally, f and the divisor $min_{a,K}$ of f are fully separable.⋄

Definitions and remarks 4 Let A be an associative K-algebra over a field K. A subalgebra T of A is called a torus of A, if T is commutative and every element of T is fully separable over K. T is maximal, if T is not contained in a proper torus of A.

A torus of a restricted Lie algebra $(L;[p])$ is an abelian restricted subalgebra of L such that every element is fully separable.

A torus of an associative algebra is a torus of the associated restricted Lie algebra (in case of positive characteristic p). The following example shows us that the reverse implication is not correct.⋄

Example 3 *(Siciliano)* Let K be a field with three elements and $L := gl(2,K)$. We focus on the element $X = \begin{pmatrix} 1 & 0 \\ 0 & -1 \end{pmatrix}$. $T = \{0, X, -X\}$ is a torus of $(L;[p])$, but T is not an associative subalgebra of $K^{2\times 2}$. For maximal tori the situation is different which is proven in the next proposition.⋄

The first part of the following proposition is included in an article of Salvatore Siciliano for which an alternative proof is given now:

Proposition 10 *Let A be a finite-dimensional associative unitary K-algebra over a field K and $S \subseteq A$.*

(i) *If S is consisting of pairwise commuting fully separable elements, then $\langle S \rangle_A$ is a torus of A (Siciliano).*

(ii) *The maximal tori of A and A° are identically. In particular, maximal tori of A° are associative subalgebras of A.*

(iii) *If $\operatorname{char} K = p > 0$ and $[p]$ is the p-function $x \mapsto x^p$, then the maximal tori of A and of $(A^\circ;[p])$ are identically. In particular, the maximal tori of $(A^\circ;[p])$ are associative subalgebras of A.*

Proof. ad(i): This is one main result in [74], chapter 5: The set of fully separable elements in a commutative associative finite-dimensional algebra is a subalgebra. This result is applied to $\langle S \rangle_A$.

ad(ii)+(iii): By using part (i) the maximal tori of the Lie algebras are associative algebras. Thus parts (ii) and (iii) are a consequence of this statement.⋄

5.2.2 Maximal tori of group algebras based on dihedral and quaternion groups

We finish this section with an example for determining all maximal tori. This example will be enhanced within the section about solvable algebras. We focus on solvable group algebras based on groups of order $2p^n$ and fields K with $char(K) = p$. By $\mathbb{P}$ we denote the set of prime numbers.

Corollary 10 Let $p \in \mathbb{P}$, K a field with $char(K) = p \geq 3$, $n \in \mathbb{N}$ and G a group of order $2 \cdot p^n$. If H is a 2-Sylow subgroup, then the conjugates under $1 + rad(KG)$ of KH are exactly the maximal tori of KG. In particular, all maximal tori of KG are of dimension 2.

Proof. The p-Sylow subgroup is normal, of order p^n and of index 2. Thus the complement H in preliminary remark 1 is of order 2. In addition, again by the preliminary remark 1 the subalgebra KH is a commutative separable radical complement in the solvable algebra KG. If t is a fully separable element, then – using results in chapter 5 of [74] – the subalgebra generated by T is separable. Hence this subalgebra is (using the enhanced theorem of Wedderburn-Malcev in [74]) contained modulo conjugation with $1 + rad(KG)$ in KH. The proof is finished.◇

By specializing the previous corollary we deduce:

Corollary 11 Let $p \in \mathbb{P}$, K a field with $char(K) = p \geq 3$, $n \in \mathbb{N}$, $G = D_{2p^n}$ and a an involution in G. All maximal tori of KG are of dimension 2. These are exactly the conjugates under $1 + rad(KG)$ of $K\langle a \rangle_{\mathcal{G}}$.[1]◇

Based on a similar argumentation we derive the following corollary (see exercises):

Corollary 12 Let $p \in \mathbb{P}$, K a field with $char(K) = p \geq 3$, $n \in \mathbb{N}_{\geq 2}$, $G = Q_{4p^n}$ and b an element of order 4 in G. All maximal tori of KG are of dimension 4. These are exactly the conjugates under $1 + rad(KG)$ of $K\langle b \rangle_{\mathcal{G}}$.◇

5.2.3 Cartan subalgebras

Definition and remark 1 Let V be a finite-dimensional vector space over a field K of characteristic 0. A Lie algebra $L \subseteq \text{End}_K(V)$ (K-linear endomorphism of V) is called splittable if for every element $f \in L$ the fully separable and nilpotent part of the generalized Jordan decomposition of f is contained in L.◇

The proof of the following proposition – except for part (v) – is straightforward to execute. Part (v) is a consequence of results of chapter 5 in [74]. Thus the proof is an exercise for the reader and not given here:

Proposition 11 Let A be an associative K-algebra, H a subset and α a monomorphism of A.

(i) α is a monomorphism of A° and $A\alpha$ is an associative subalgebra of A.

---

[1] Let G be a group and T a subset of G. By $\langle T \rangle_{\mathcal{G}}$ we denote the subgroup generated by T in G.

(ii) $C_{A\alpha}(H\alpha) = C_A(H)\alpha$

(iii) H is a Cartan subalgebra of A° if and only if $H\alpha$ is a Cartan subalgebra of $(A\alpha)^\circ$.

(iv) H is a (maximal) torus of A° if and only if $H\alpha$ is a (maximal) torus of $(A\alpha)^\circ$.

(v) Let K be a field and $a \in A$. If a possesses a generalized Jordan decomposition $(v; n)$, then v and n are polynomials in a. In particular, the Lie algebra $(A\rho)^\circ$ is splittable if K is a perfect field.◊

The following result is contained in [10] (section VII.5, proposition 6) and is the basis result for determining the Cartan subalgebras of associated Lie algebras:

Lemma 3 *Let V be a finite-dimensional vector space over a field K with $char(K) = 0$ and L a splittable Lie subalgebra of $End_K(V)^\circ$.*

(i) *If H is a Cartan subalgebra of L, then $T = \{f \in H \mid f \text{ fully separable}\}$ is a maximal torus of L.*

(ii) *If T is a maximal torus of L, then $C_L(T)$ is a Cartan subalgebra of L.*◊

Based on lemma 3 we proof now the following theorem of Salvatore Siciliano. The idea within the proof is likewise to the one included in [59] and revised.

Theorem 20 *(Siciliano) Let A be an associative unitary K-algebra of finite dimension over a field K and H a subset of A. H is a Cartan subalgebra of A° if and only if a maximal torus T of A exists such that $H = C_A(T)$ is valid.*

Proof. In the case $\mathrm{char}\, K = p > 0$ the result can be deduced by proposition 10 and a classical result within the theory of restricted Lie algebras (see [65], theorem 4.1 in chapter 2).
Now let K be of characteristic zero. Because of proposition 11 the Lie algebra $(A\rho)^\circ$ is splittable.
Let H be a Cartan subalgebra of A°. Proposition 11 implies that $H\rho$ is a Cartan subalgebra of $(A\rho)^\circ$. By using lemma 3 we deduce that the set T of fully separable elements of $H\rho$ is a maximal torus of $(A\rho)^\circ$. Proposition 10 is used to prove that its a maximal torus of $A\rho$. By using corollary 9 we get that T is central in $H\rho$, and hence $H\rho$ is contained in $C_{A\rho}(T)$. This centralizer is with respect to lemma 3 a Cartan subalgebra, and by the maximal nilpotency we conclude $H\rho = C_{A\rho}(T)$. Proposition 11 implies that H is exactly the centralizer of the maximal torus – the pre-image of T under

ρ.

Now let $H = C_A(T)$ be valid for a maximal torus of A. Proposition 10 implies that $T\rho$ is a maximal torus of $(A\rho)^\circ$. By using lemma 3 we deduce that $C_{(A\rho)^\circ}(T\rho)$ is a Cartan subalgebra of $(A\rho)^\circ$. Finally, proposition 11 and the identity $C_{(A\rho)^\circ}(T\rho) = C_A(T)\rho$ prove that $C_A(T)$ is a Cartan subalgebra of A°. ◇

The proof of theorem 20 (in the case $char(K) = 0$), of proposition 10 and of theorem 4.1 in chapter 2 of [65] (in the case of $char(K) > 0$) yields to the following main theorem:

Main theorem 1 *Let A be an associative unitary K-algebra of finite dimension over a field K, $\mathcal{T}_A$ the set of maximal tori of A and $\mathcal{C}_{A^\circ}$ the set of Cartan subalgebras of A°. The functions*

$$C_A(\cdot) : \mathcal{T}_A \longrightarrow \mathcal{C}_{A^\circ}, T \mapsto C_A(T)$$

and

$$VSEP(\cdot) : \mathcal{C}_{A^\circ} \longrightarrow \mathcal{T}_A, C \mapsto \{v \mid v \in C, v \text{ is fully separable}\}$$

are inverse (bijections) to each other. In particuluar, Cartan subalgebras are existing. ◇

In particular, under the assumptions of the main theorem 1 Cartan subalgebras exists. This existence can be proven also without the main theorem by using classical results for Lie algebras in characteristic zero – in this case Cartan subalgebras are existing because they are minimal Engel subalgebras – and for restricted Lie algebras – in this case Cartan subalgebras are also the centralizers of the maximal tori (as used within the proof of the theorem).

5.2.4 Cartan subalgebras of group algebras based on dihedral and quaternion groups

Within this section we determine Cartan subalgebras of $(KG)^\circ$ for dihedral and quaternion groups (D_{2p^n} and Q_{4p^n}) and a field of characteristic p. By using our main theorem 1 and corollary 10 we deduce directly:

Corollary 13 *Let $p \in \mathbb{P}$, K a field of characteristic $p \geq 3$, $n \in \mathbb{N}$ and G a group of order $2 \cdot p^n$. If H is a 2-Sylow subgroup, then the conjugates under $1 + rad(KG)$ of $C_{KG}(KH)$ are the Cartan subalgebras of $(KG)^\circ$.* ◇

As a consequence we deduce for dihedral groups:

Corollary 14 *Let $p \in \mathbb{P}$, K a field with characteristic $p \geq 3$, $n \in \mathbb{N}$, $G = D_{2p^n}$ and a an involution of G. The Cartan subalgebras of $(KG)^\circ$ are the conjugates under $1 + rad(KG)$ of the centralizer of $K\langle a \rangle_{\mathcal{G}}$ in KG.* ◇

Within section 5.5.4 we will calculate this centralizer and its dimension explicitly. The dimension is identical for all Cartan subalgebras, and it is $p^n + 1$. For the quaternion groups the following results (based on a similar argumentation) is valid:

Corollary 15 *Let $p \in \mathbb{P}$, K a field with characteristic $p \geq 3$, $n \in \mathbb{N}_{\geq 2}$, $G = Q_{4p^n}$ and b an element of order 4 in G. The Cartan subalgebras of $(KG)^\circ$ are the conjugates under $1 + rad(KG)$ of the centralizer of $K\langle b\rangle_\mathfrak{g}$ in KG.* ◇

We remark that the centralizer is of dimension $2 \cdot (p^n + 1)$ which we will prove in section 5.5.4, too.

5.2.5 Maximal tori and Cartan subalgebras of Solomon algebras, Solomon-Tits algebras and triangular matrices

A first introduction to these algebras is included in part of section 4.2.2 (and the following sections, too). All these algebras are solvable, and these kind of algebras will be analyzed in an own section, too. For demonstrating our results the author has included this section already at this point. In [76] it is proven that all radical complements of the Solomon-Tits algebras are self-centralizing. Thorsten Bauer proves in his dissertation the same result for the Solomon algebras in characteristic zero (see [4]). In addition, it is straightforward to prove that the subalgebra of diagonal matrices is a self-centralizing radical complement for the algebra of lower and upper triangular matrices.

Let A be an associative unitary solvable K-algebra with a separable factor algebra by its nilradical and a self-centralizing radical complement in A. We will prove later on that under these assumptions the radical complements coincide with the maximal tori of A. For this it is not necessary that the radical complements are self-centralizing. But in this case – by using the main theorem 1 – they are identical to the Cartan subalgebras of A°.

The main idea for the proof that maximal tori and radical complements are identical is the following: in a first step it is proven that tori are commutative separable subalgebras. By using the theorem of Wedderburn-Malcev they are contained in a radical complement. Finally, we have to deduce that the radical complements are indeed tori. But this is more or less true by our assumption that A is solvable and possesses a separable factor algebra by its nilradical.

For the algebra A lower and upper triangular matrices over a field K the set of diagonal matrices - denoted by $D(n, K)$ - is a radical complement. All radical complements are conjugated under $1 + rad(A)$ of this complement, and they are of dimension n.

$K\Pi_n$ - the Solomon-Tits algebra (see e.g. [76]) - possesses a self-centralizing radical complement of dimension $B(n)$ - the so-called Bell numbers of n. Descriptions and results for this statement can be found e.g. in [76]. All radical complements are conjugated under $1+rad(K\Pi_n)$ to this complement.

D_n - the Solomon algebra in $char(K) = 0$ (see e.g. [4]) - possesses a self-centralizing radical complement of dimension $p(n)$ - the so-called partition numbers of n. Descriptions and results for this statement can be found e.g. in [76]. All radical complements are conjugated under $1 + rad(D_n)$ to this complement.

Section 5.2

{ unusual topic of A }

{ Certain subalgebras of A }

$\mathcal{J}A$ → $C_A(\cdot)$

$VSer(\cdot)$ ↔ 1:1 ← $C_A(\cdot)$

$VSer(C_A(\cdot))$

$C = C_A(VSer(\cdot))$

$T = VSer(C_A(T))$

C_{10}

- Solomon algebra
- Solomon-Tits algebra } reduced complements are self-centralizing
- lower and upper triangular matrices

- $u D_{2,n} u D_{2,n}$ ... not ... ⟺ $C_{10} \neq \mathcal{J}A$
  $Dim = ?$ → role in $CL.6$!

5.2.6 Open-ended questions and exercises

Open-ended questions 3 *(i) Is it possible to transfer the results of this section to infinite-dimensional or right Artinian algebras?*

(ii) What are the maximal abelian subalgebras of a Lie algebra associated to an associative algebra?

(iii) What are the maximal nilpotent subalgebras of a Lie algebra associated to an associative algebra?

(iv) What are the maximal solvable subalgebras of a Lie algebra associated to an associative algebra?

(v) What are the solutions of these questions for the Jordan algebra associated to an associative algebra?

Excercise 89 *Let A be an associative unitary K-algebra and D a semisimple subalgebra of A. $1+rad(A)$ is acting on the set of semisimple subalgebras by conjugation. The stabilizer of D with respect to this action is the normalizer in $1 + rad(A)$ of D and the identity $N_{1+rad(A)}(D) = C_{1+rad(A)}(D)$ is valid. What is the meaning of this identity for a radical complement? Deduce a formula for the quantity of radical complements in the case of a finite field. What is the consequence for an algebra possessing a self-centralizing radical complement?*

Excercise 90 *Use exercise 89 to calculate the number of Cartan subalgebras and maximal tori of A° for the following cases:*

(i) $A := K\Pi_n$ the Solomon-Tits algebra, K a finite field, $n \in \mathbb{N}$ (see e.g. [76])

(ii) $A = D_n$ the Solomon algebra, K a field of characteristic zero, $n \in \mathbb{N}$ (see e.g. [76])

(iii) $A =$ The algebra lower triangular matrices of $K^{n \times n}$, K a finite field, $n \in \mathbb{N}$

(iv) $A =$ The algebra of upper triangular matrices of $K^{n \times n}$, K a finite field, $n \in \mathbb{N}$

(v) $A = KG$, $p \in \mathbb{P}$, $n \in \mathbb{N}$, K a finite field of characteristic $p \geq 3$ and G a dihedral resp. quaternion group of order $2p^n$ resp. $4p^n$.

Excercise 91 *Prove all statements with definition and remark 3!*

Excercise 92 *Every divisor of a semisimple and separable polynomial is semisimple and separable, too! Is this true for nilpotency as well?*

Excercise 93 *Prove proposition 11!*

Excercise 94 *Prove example 3!*

Excercise 95 *Prove corollary 12!*

Excercise 96 *Prove corollary 15!*

Excercise 97 *Prove main theorem 1 exactly!*

Excercise 98 *What is the exact meaning of main theorem 1? In what way are all Cartan subalgebras and maximal tori are determined and connected to each other? Composite the functions $C_A(\cdot)$ and $VSEP(\cdot)$ in two possible variants and derive two equations for maximal tori and Cartan subalgebras!*

Excercise 99 *Let A be a finite-dimensional associative unitary K-algebra over a finite field K. The sets of maximal tori and Cartan subalgebras of A° are of equal order. Specify a bijection between these two sets!*

Excercise 100 *Let K be a finite field and $n \in \mathbb{N}$. Is every Lie subalgebra of $gl(n, K)$ splittable? If the answer is negative, then create an example with preferably small values for n and K such that $gl(n, K)$ is non-splittable.*

5.3 Division algebras

Salvatore Siciliano proves within [59], theorem 2, that the separable maximal subfields of a central finite-dimensional associative K-division algebra are exactly the Cartan subalgebras of its associated Lie algebra. This theorem will be proven based on a different idea within this section. In addition, we enhance the theorem to non-central division algebras. Within this section we prove that the maximal separable and separable maximal subfields are identical for central division algebras. A consequence is an alternative prove of a theorem of Emmy Noether[2].

---

[2]Amalie Emmy Noether (born 23 March 1882, died 14 April 1935) was a German Jewish mathematician known for her landmark contributions to abstract algebra and theoretical physics. She was described by Pavel Alexandrov, Albert Einstein, Jean Dieudonné, Hermann Weyl, and Norbert Wiener as the most important woman in the history of mathematics. As one of the leading mathematicians of her time, she developed the theories of rings, fields, and algebras. In physics, Noether's theorem explains the connection between symmetry and conservation laws. Noether was born to a Jewish family in the Franconian town of Erlangen; her father was a mathematician, Max Noether. She originally planned to teach French and English after passing the required examinations, but instead studied mathematics at the University of Erlangen, where her father lectured. After completing her dissertation in 1907 under the supervision of Paul Gordan, she worked at the Mathematical Institute of Erlangen without pay for seven years. At the time, women were largely excluded from academic positions. In 1915, she was invited by David Hilbert and

5.3.1 Central division algebras

Proposition 12 *Let D be a finite-dimensional associative non-commutative K-division algebra. Then D° is not nilpotent.*

Proof. D is as $Z(D)$-algebra a central division algebra, and D° is nilpotent as K-algebra if and only if its nilpotent as $Z(D)$-algebra. Hence we assume that D is central.
Let T be a maximal subfield of D, and we assume that D is Lie nilpotent. Hence $(D \otimes T)^\circ$ is nilpotent, too (see remark 8). It is well-known (see e.g. [49]) that $D \otimes T$ is isomorphic to $T^{n \times n}$ with $n = ind(D)$. By using remark 9 we get $n = 1$ and D is nilpotent. $\diamond$

We enhance theorem 2 in [59] by proving the following theorem:

Theorem 21 *Let D be a central finite-dimensional associative K-division algebra.*

(i) The maximal separable subfields are exactly the separable maximal subfields of D. (MAXSEP=SEPMAX)

(ii) A separable maximal subfield exists. (E. Noether)

Felix Klein to join the mathematics department at the University of Göttingen, a world-renowned center of mathematical research. The philosophical faculty objected, however, and she spent four years lecturing under Hilbert's name. Her habilitation was approved in 1919, allowing her to obtain the rank of Privatdozent. Noether remained a leading member of the Göttingen mathematics department until 1933; her students were sometimes called the "Noether boys". In 1924, Dutch mathematician B. L. van der Waerden joined her circle and soon became the leading expositor of Noether's ideas: her work was the foundation for the second volume of his influential 1931 textbook, Moderne Algebra. By the time of her plenary address at the 1932 International Congress of Mathematicians in Zürich, her algebraic acumen was recognized around the world. The following year, Germany's Nazi government dismissed Jews from university positions, and Noether moved to the United States to take up a position at Bryn Mawr College in Pennsylvania. In 1935 she underwent surgery for an ovarian cyst and, despite signs of a recovery, died four days later at the age of 53. Noether's mathematical work has been divided into three "epochs". In the first (1908 to 1919), she made contributions to the theories of algebraic invariants and number fields. Her work on differential invariants in the calculus of variations, Noether's theorem, has been called "one of the most important mathematical theorems ever proved in guiding the development of modern physics". In the second epoch (1920 to 1926), she began work that "changed the face of abstract algebra". In her classic paper Idealtheorie in Ringbereichen (Theory of Ideals in Ring Domains, 1921) Noether developed the theory of ideals in commutative rings into a tool with wide-ranging applications. She made elegant use of the ascending chain condition, and objects satisfying it are named Noetherian in her honor. In the third epoch (1927 to 1935), she published works on noncommutative algebras and hypercomplex numbers and united the representation theory of groups with the theory of modules and ideals. In addition to her own publications, Noether was generous with her ideas and is credited with several lines of research published by other mathematicians, even in fields far removed from her main work, such as algebraic topology.

(iii) The Cartan subalgebras of D° are exactly the separable maximal subfields of D. (S. Siciliano)

Proof. ad(i): Let T be a maximal separable subfield of D. Then T is a maximal torus of D, because every unitary subalgebra of D is a division algebra and tori are commutative. By using the main theorem 1 we deduce that $C_D(T)$ is a Cartan subalgebra of D°. D is a finite-dimensional associative K-division algebra, and hence $C_D(T)$ has the same properties. Proposition 12 implies that $C_D(T)$ is a subfield of D. By using the maximal Lie nilpotency of $C_D(T)$ we conclude that $C_D(T)$ is a maximal subfield of D. Maximal subfields are self-centralizing (see e.g. [49]), and thus $C_D(C_D(T)) = C_D(T)$ is valid. The double-centralizer theorem implies $C_D(C_D(T)) = T$. Hence $T = C_D(T)$ is a separable maximal subfield of D. The other implication is straightforward to prove.

ad(ii): $K1_D$ is a separable subfield of the finite-dimensional division algebra D. Hence a maximal separable subfield of D exists. By using part (i) we prove part (ii).

ad(iii): By using the main theorem 1 the Cartan subalgebras of D° are exactly the centralizers of the maximal tori of D. A maximal torus of D is a maximal separable subfield of D, because every unitary subalgebra of D is a K-division algebra and tori are commutative. Part (i) implies that the Cartan subalgebras of D° are exactly the centralizers of the separable maximal subfields. Its well-known that (see e.g. [49]) every maximal subfield of D is self-centralizing. ⋄

A direct consequence of theorem 21 is the following corollary:

Corollary 16 *Let D be a central associative finite-dimensional K-division algebra.*

(i) All Cartan subalgebras of D° are isomorphic and $\mathrm{ind}(D)$-dimensional.

(ii) If K is perfect, then the Cartan subalgebras of D° are exactly the maximal subfields of D.

(iii) The maximal tori of D are exactly the Cartan subalgebras of D°. ⋄

5.3.2 Non-central division algebras

In this section we enhance theorem 21 to finite-dimensional associative – not necessarily central – K-division algebras. For this its purposeful to focus on K-division algebras as division algebra over its center $Z(D)$. As $Z(D)$-algebra D is central.

83

Theorem 22 *Let D be a finite-dimensional associative K-division algebra.*

(i) The Cartan subalgebras of D° are exactly the maximal subfields of D which are separable over $Z(D)$.

(ii) A maximal subfield exists which is separable over $Z(D)$.

(iii) The maximal subfields of D which separable over $Z(D)$ are exactly the subfields of D which are maximal separable over $Z(D)$.

Proof. We remark that every maximal subfield D is containing the center of D because its self-centralizing.

ad(i): Let T be a maximal subfield of D being separable over $Z(D)$. D is central as $Z(D)$-algebra. By using theorem 21 the field T is a Cartan subalgebra of D° as $Z(D)$-algebra. Obviously, T is a Cartan subalgebra of D° as K-algebra, too: nilpotency and self-normality are independent of the base field.
Let H be a Cartan subalgebra of D° as K-algebra. Because of $Z(D) \leq N_{D^\circ}(H) = H$ and proposition 8 is the algebra H a Cartan subalgebra of D° as $Z(D)$-algebra. Theorem 21 implies part (i).

ad(ii)+(iii): Focus on D as $Z(D)$-algebra and use theorem 21. ⋄

A direct consequence of theorem 22 is the following corollary:

Corollary 17 *Let D be a finite-dimensional associative K-division algebra.*

(i) All Cartan subalgebras of D° are isomorphic and of dimension $\mathrm{ind}_{Z(D)}(D) \cdot \dim_K(Z(D))$.

(ii) If $Z(D)$ is separable over K, then the Cartan subalgebras of D° are exactly the separable maximal subfields of D.

Add-on: D is as algebra separable if and only if $Z(D)$ is separable over K. In this case maximal tori and Cartan subalgebras coincide.

(iii) If K is perfect, then the Cartan subalgebras of D° are exactly the maximal subfields of D. ⋄

Section 5.3

(cartoon cloud: Cartier division algebras)

ID
 maximal torus
 ∥
 maximal separable
 subfield
   → self-centralizing
   → maximal commutative subalgebras

E/v°
 Cartan subalgebras
   { abelian and
     ind(v)-
     diagonal
   → $C_v(·)$ and $WE+(·)$
     are the identity
     functor
     ∥
     Cartan
     subalgebras
   { abelian and
     ind(v)•dim(R(V))-
     diagonal

max mal torus
 ∥
 maximal separable
 sub $Z(G)$
 ∥
 separable over $Z(G)$ and
 maximal subfield

(cartoon cloud: division algebras)

$\mathbb{Z}_p$ torsors

5.3.3 Open-ended questions and exercises

Open-ended questions 4 *(i) Is it possible to transfer the results of this section to infinite-dimensional or right Artinian algebras?*

(ii) What are the solutions of this section for the Jordan algebra associated to an associative algebra?

Let D be a finite-dimensional associative K-division algebra.

Excercise 101 *Prove that every maximal subfield of D is self-centralizing.*

Excercise 102 *True or false: If the Cartan subalgebras of D° are exactly the maximal subfields of D, then K is perfect.*

Excercise 103 *Prove that every maximal subfield of D is containing the center of D.*

Excercise 104 *Prove that every unital subalgebra of D is a K-division algebra. What is the consequence for $Z(D)$?*

Excercise 105 *Is every subalgebra of D unital?*

Excercise 106 *Prove that D is a $Z(D)$-algebra. Is D also a T-algebra for every unital subalgebra T of D contained in the center of D?*

Excercise 107 *Prove that D° is as $Z(D)$-algebra nilpotent resp. solvable if and only if it is nilpotent resp. solvable as K-algebra.*

Excercise 108 *Prove that every Cartan subalgebra of D° is containing the center of D.*

Excercise 109 *Which connection exists between the maximal tori and the Cartan subalgebras of D°?*

Excercise 110 *Which connection exists between the maximal tori and the Cartan subalgebras of D° as K- and as $Z(D)$-algebra?*

Excercise 111 *What is the dimension of the Cartan subalgebras and maximal tori of D° in the case that D is four-dimensional?*

Excercise 112 *(maximal tori = Cartan subalgebras) Prove that the maximal tori and the Cartan subalgebras are identical if and only if D is a separable K-algebra. This is valid if and only if every maximal torus is self-centralizing. A sufficient condition for this statement is that K is perfect.*

5.4 Quaternion algebras

5.4.1 The case characteristic $\neq 2$

Let K be a field with $char(K) \neq 2$ and D a four-dimensional associative central K-division algebra. Then its well-known that there exist $a, b \in K \setminus \{0\}$ and a K-basis $\{1, i, j, k\}$ which define a multiplication by:

·	1	i	j	k
1	1	i	j	k
i	i	$a1$	k	aj
j	j	-k	$b1$	-bi
k	k	-aj	bi	$-ab1$

These division algebras are called (generalized) quaternion algebras. We determine the separable maximal subfields of D. The index $ind(D)$ of D is equal to 2. Thus all maximal subfields of D are two-dimensional and are containing $K1_D$. We have to calculate the separable elements $t \in D \setminus K1_D$ such that $t^2 \in \langle 1, t \rangle_K$ is valid.

Let $k_1, k_2, k_3 \in K$ and $t = k_1 i + k_2 j + k_3 k$. A straightforward calculation shows $t^2 = (k_1^2 a + k_2^2 b - k_3^2 ab)1 \in \langle 1, t \rangle_K$. Thus we deduce $min_{t,K} = x^2 - t^2 = (x+t)(x-t)$. Because of $char(K) \neq 2$ the element t is separable over K, and $\langle 1, t \rangle_K$ is a separable maximal subfield. Hence every maximal subfield is separable.◇

5.4.2 Examples for the associative conjugacy

(i) In the real quaternion algebra $\mathbb{H}$ all maximal subfields are separable over $\mathbb{R}$ and two-dimensional field extensions of $\mathbb{R}$. Hence they are isomorphic to $\mathbb{C}$, and by a theorem of Skolem-Noether they are conjugated in $\mathbb{H}$.

(ii) We focus on the rational quaternion algebra D in the case $a = -1$ and $b = -2$, and we prove that $K[i]$ and $K[j]$ are non-isomorphic maximal subfields of D. We assume the contrary. By using a theorem of Skolem-Noether a unit g would exist, like $g = k_0 1 + k_1 i + k_2 j + k_3 k$, with $i^g \in K[j]$. By [49] the inverse is to be calculated by $g^{-1} = v(g)^{-1} g^\star$, for which $v(g)$ is an element of $K \setminus \{0\}$ and $g^\star$ is the conjugate of g: $g^\star = k_0 1 - k_1 i - k_2 j - k_3 k$ and $v(g) = g \cdot g^\star = k_0^2 + k_1^2 + k_2^2 + k_3^2$. Comparing coefficients would imply:

$$(1) \quad k_2 k_0 + k_1 k_3 = 0$$
and
$$(2) \quad k_0^2 + k_1^2 - 2k_2^2 - 2k_3^2 = 0.$$

Let $k_2 = 0$. Then we would deduce by (1) the condition $k_1 = 0$ or $k_3 = 0$, and by (2) we would get $g = 0$ or $\sqrt{2} \in \mathbb{Q}$.

Let $k_2 \neq 0$. By using (1) and (2) we would calculate $\frac{k_1^2 k_3^2}{k_2^2} + k_1^2 - 2k_2^2 - 2k_3^2 = 0$, hence $k_1^2(k_2^2 + k_3^2) - 2k_2^2(k_2^2 + k_3^2) = 0$, also $k_1^2 = 2k_2^2$ would be valid and again we would deduce $\sqrt{2} \in \mathbb{Q}$.⋄

5.4.3 Remark on the Lie conjugacy

Let A be an associative K-algebra and $a \in A$. By ρ_a resp. λ_a we denote the right resp. left multiplication with a in A.

Let K be an algebraically closed field of characteristic 0 and L a finite-dimensional K-Lie algebra. Its well-known that all Cartan subalgebras of L are conjugated under the inner group of automorphism of L. This group consists of the elements $exp(ad(l))$ such that $l \in L$ and $ad(l)$ is nilpotent. The nilpotency of $ad(l)$ ensures that the exponential series can be calculated at $ad(l)$.

In contrast to this result we obtain: Let K a field with $char(K) = 0$ and D a finite-dimensional associative K-division algebra. By corollary 17 all Cartan subalgebras of D° are exactly the maximal subfields of D. If $l \in D$, then l is – by using $char(K) = 0$ – separable over K. ρ_l and λ_l possess the same minimal polynomial as l and commute. By using 5.3.1 in [74] we deduce that $ad(l) = \rho_l - \lambda_l$ is separable over K. In particular, $ad(l)$ is nilpotent if and only if $ad(l) = 0$ is valid. This statement is equivalent to l being central. In this case $exp(ad(l)) = id_D$ is valid.⋄

5.4.4 The case characteristic 2

Let K a field with $char(K) = 2$ and D a four-dimensional associative central K-division algebra. It is well-known that there exist $a, b \in K \setminus \{0\}$ and a K-basis $\{1, i, j, k\}$ with the following multiplication:

·	1	i	j	k
1	1	i	j	k
i	i	a1	k	aj
j	j	k+i	j+b1	b1
k	k	a(j+1)	k+bi	ab1.

These division algebras are called (generalized) quaternion algebras. We determine the separable maximal subfields of D. The index $ind(D)$ is 2 and hence all maximal subfields of D are two-dimensional and are containing $K1_D$. Thus we have to calculate the separable elements $t \in D \setminus K1_D$ such that $t^2 \in \langle 1, t \rangle_K$ is valid. Let $k_1, k_2, k_3 \in K$ and $t = k_1 i + k_2 j + k_3 k$.

<u>Case 1:</u> $k_2 = k_3 = 0$:
Then $k_1 \neq 0$ is valid, and we deduce $T = K[i]$. By $i^2 = a1$ we calculate the minimal polynomial $min_{i,K} = x^2 + a = (x+i)^2$. Thus T is a non-separable

but maximal subfield of D.

<u>Case 2:</u> $k_3 = 0, k_2 \neq 0$:
We deduce $t = k_1 i + k_2 j$, and a calculation shows us $t^2 = (k_1^2 a + k_2^2 b)1 + k_2 t \in \langle 1, t \rangle_K$. This implies $min_{t,K} = x^2 + k_2 x + (k_1^2 a + k_2^2 b) = (x+t)(x+t+k_2)$. Thus T is a separable maximal subfield of D.

<u>Case 3:</u> $k_3 \neq 0, k_2 = 0$:
$t = k_1 i + k_3 k$ and $t^2 = (k_1^2 a + k_3^2 ab + k_3 k_1 a)1$ are valid. By this we deduce $min_{t,K} = (x+t)^2$, and hence T is a non-separable but maximal subfield of D.

<u>Case 4:</u> $k_3 k_2 \neq 0$:
Its straightforward to prove that $t^2 \in \langle 1, t \rangle_K$ is only for $k_3 k_2 bi \in \langle 1, t \rangle_K$ valid. This statement is equivalent to $\langle 1, t \rangle_K = K[i]$, and by using case 1 the field $\langle 1, t \rangle_K$ is a non-separable but maximal subfield of D.

In summarizing, only in case 2 a separable maximal subfield is existing. Hence the Cartan subalgebras are exactly $\{\langle 1, t \rangle_K \mid \exists k_1, k_2 \in K : k_2 \neq 0, t = k_1 i + k_2 j\}$.⋄

Section 5.4

```
        ┌─────────────┐        ┌──────────────┐        ┌──────────────┐
        │  C̄ ar ≠ 2   │        │  Quaternion  │        │   C̄ ar = 2   │
        └─────────────┘        │   algebras   │        └──────────────┘
                               └──────────────┘
```

C̄ ar ≠ 2

$J_0 = C_\infty$

every unital
satisfied in separate
and 2-dimensional

↳ $k := \mathbb{R}$, all are conjugated
$k := \mathbb{Q}$, $a=\sqrt{2}$, $b=\sqrt{3}$, $\mathbb{Q}[i]$ and $\mathbb{Q}[\sqrt{2}]$ are not conjugated

```
●────────●────────●────────●
0        1        2       A(a,b)
a       u₁=1      T
```

Quaternion algebras

↳ central division algebra

C̄ ar = 2

$J_0 = C_\infty$

$\langle 1, t \rangle u$ with
$t = k_1 \cdot i + k_2 \cdot j$ and
$k_1 \in k$, $0 \neq i \in k$
are the separable unital
subjects

```
●────────●────────●────────●
0        1        2       A(a,b)
         u₁=1
         T
```

5.4.5 Open-ended questions and exercises

Open-ended question 1 *Are there other families of (central) division algebras and what are their separable maximal subfields?*

Excercise 113 *Which connection exists between the quaternion group Q_8 and an arbitrary quaternion algebra?*

Excercise 114 *Prove the identity $g^{-1} = v(g)^{-1}g^\star$ (which was used in this section) for an element g in an arbitrary quaternion algebra. Deduce under what conditions g is invertible.*

Excercise 115 *The quaternion conjugation $\star$ is an anti-automorphism of every quaternion algebra!*

Excercise 116 *Determine the maximal subfields of all complex quaternion algebras. Is there a conjecture for arbitrary algebraically closed fields?*

Excercise 117 *Determine the maximal subfields of all real quaternion algebras.*

Excercise 118 *What are the maximal subfields of the rational quaternion algebra in the case $a = -1$ and $b = -2$? Is it possible to determine them for arbitrary a, b?*

Excercise 119 *What are the maximal subfields of the quaternion algebras over a finite field with p elements (p a prime number)? Is there a conjecture for arbitrary finite fields?*

Excercise 120 *Determine all unital subalgebras of quaternion algebra in characteristic 2. Is it possible to describe the non-unital ones, too?*

Excercise 121 *Determine all unital subalgebras of quaternion algebra in characteristic $\neq 2$. Is it possible to describe the non-unital ones, too?*

5.5 Solvable algebras

In [4] and [59] Thorsten Bauer and Salvatore Siciliano proved that for a finite-dimensional associative unitary solvable K-algebra A with separable factor algebra by its nilradical the Cartan subalgebras of A° are exactly the centralizers of the radical complements of A.

We will prove this result based on a different approach, enhance it to non-unitary algebras and focus on solvable group algebras in general and for dihedral and quaternion groups. We finish this section by analyzing solvable associative algebras for which the base field is a splitting field. In this case, Cartan subalgebras are related to Pierce components.

5.5.1 Unitary solvable algebras

Our analysis is based on the following lemma (see e.g. theorem 5.3.1 in [74]):

Lemma 4 *Let A be a finite-dimensional associative commutative unitary K-algebra. A is separable if and only if every element of A is fully separable over K.*⋄

Theorem 23 *Let A be a finite-dimensional associative unitary solvable K-algebra with separable factor algebra by its nilradical. The maximal tori of A are exactly the radical complements of A.*

Proof. '→:' Let T be a radical complement of A. Then T is a commutative separable unitary subalgebra of A, and with respect to lemma 4 its a torus of A. Let S be a torus of A containing T. Again using lemma 4 the subalgebra S is a separable K-subalgebra of A, and thus it is direct to $rad(A)$. Comparing dimensions we deduce $T = S$.
'←:' Let T be a maximal torus of A. By using lemma 4 the subalgebra T of A is separable. An enhanced version of the conjugacy theorem of Wedderburn-Malcev (see e.g. corollary 2.3.7 in [74]) implies that T is contained in a radical complement of A. By the proof of '→:' we deduce that this complement is a torus of A. By the maximality of T it coincides with this complement.⋄

By using theorem 23, theorem 1 in [59] and the theorem of Wedderburn-Malcev we deduce the following theorem:

Theorem 24 *Let A be a finite-dimensional associative unitary solvable K-algebra with separable factor algebra by its nilradical.*

(i) The Cartan subalgebras of A° are exactly the centralizers of the radical complements of A. (T. Bauer)

(ii) All Cartan subalgebras of A° are conjugated under the normal nilpotent subgroup $1_A + rad(A)$ of the group of units of A. (T. Bauer)

(iii) The maximal tori of A coincide with the Cartan subalgebras of A° if and only if a self-centralizing radical complement exists.

(iv) The maximal tori of A coincide with the Cartan subalgebras of A° if and only if all radical complements are self-centralizing.⋄

Examples 1 (i) Within his dissertation [4] Thorsten Bauer proves that the radical complements of the (solvable) Solomon algebras in zero characteristic are self-centralizing. This is true for the Solomon-Tits algebras in arbitrary characteristic, too (see [76]).

(ii) Within the algebra A of lower resp. upper triangular matrices over an arbitrary field the subalgebra of diagonal matrices is a radical complement of A which is self-centralizing.

(iii) For an arbitrary field K we focus on the algebra A of lower triangular matrices in $K^{5\times 5}$. The set T of matrices of the form

$$\begin{pmatrix} a & 0 & 0 & 0 & 0 \\ 0 & b & 0 & 0 & 0 \\ 0 & 0 & 0 & 0 & 0 \\ 0 & 0 & 0 & 0 & 0 \\ 0 & e & c & d & 0 \end{pmatrix}$$

is a subalgebra of A, and its nilradical can be presented by the matrices of the form

$$\begin{pmatrix} 0 & 0 & 0 & 0 & 0 \\ 0 & 0 & 0 & 0 & 0 \\ 0 & 0 & 0 & 0 & 0 \\ 0 & 0 & 0 & 0 & 0 \\ 0 & e & c & d & 0 \end{pmatrix}$$

and one radical complement by

$$\begin{pmatrix} a & 0 & 0 & 0 & 0 \\ 0 & b & 0 & 0 & 0 \\ 0 & 0 & 0 & 0 & 0 \\ 0 & 0 & 0 & 0 & 0 \\ 0 & 0 & 0 & 0 & 0 \end{pmatrix}.$$

The centralizer of this radical complement is straightforward to be determined:

$$\begin{pmatrix} a & 0 & 0 & 0 & 0 \\ 0 & b & 0 & 0 & 0 \\ 0 & 0 & 0 & 0 & 0 \\ 0 & 0 & 0 & 0 & 0 \\ 0 & 0 & c & d & 0 \end{pmatrix}$$

is a presentation of it. In addition, it is provable that the subalgebra T is **non-unitary** and solvable (see exercises). In the next section we analyze the transfer of theorem 24 to such kind of algebras.⋄

5.5.2 The quasi regular group and non-unitary solvable associative algebras

Definition and remark 2 If A is a K-algebra, then we define $a \star b := a + b + ab$ for all $a, b \in A$. Van der Waerden calls this composition the star product on A.

If A is associative, then $(A; \star)$ is a monoid with zero element 0_A. Its group of unit $Q(A)$ is called the star group of A. For example, every nilpotent element is invertible with respect to $\star$. The inverse element of an element $r \in Q(A)$ is symbolized by $r^{(-1)}$, the conjugate element of $r \in Q(A)$ under $s \in Q(A)$ by $r^{(s)}$.

For unitary A the shifting by 1_A is a monoid isomorphism between $(A; \star)$ and $(A; \cdot)$. In particular, $Q(A)$ is isomorphic to $E(A)$. ◇

Definition and remark 3 For a K-algebra A we define a new multiplication on the K-space $K \times A$ by $(c; x)(d; y) := (cd; cy + dx + xy)$ for all $c, d, x, y \in A$. Equipped by this composition $K \times A$ is a K-algebra with identity element $(1_K; 0_A)$. This unitary K-algebra, containing A as an isomorphic copy by $\{0_K\} \times A$ is symbolized by (K, A). It is called the adjunction of A by a unit.

If $k, l \in K$, $a, b \in A$, A associative and $r \in Q(A)$, then the following identities are valid in (K, A) by which proposition 13 is a direct consequence of:[3]

$$(k; a) \circ (l; b) = (0; a \circ b)$$
$$(k; t)^{1+r} = (k; s^{(r)}). \diamond$$

Proposition 13 Let A be an associative K-algebra and T a subalgebra of A.

(i) $C_{(K,A)}((K, T)) = (K, C_A(T))$

(ii) $N_{(K,A)^\circ}((K, T)) = (K, N_{A^\circ}(T))$

(iii) T° is nilpotent if and only if $(K, T)^\circ$ is nilpotent.

(iv) T is a Cartan subalgebra of A° if and only if (K, T) is a Cartan subalgebra of $(K, A)^\circ$.

(v) If A is unitary and $r \in Q(A)$, then $(K, T)^{1+r} = (K, T^{(r)})$ is valid.

(vi) For all $r \in Q(A)$ the identity $C_A(T)^{(r)} = C_A(T^{(r)})$ is valid. ◇

Theorem 25 Let A be a finite-dimensional associative solvable K-algebra with separable factor algebra by its nilradical.

---
[3](K, A) can be used to construct an unitary algebra containing A directly. This kind of algebra was introduced earlier by A^K. Both algebras are isomorphic.

(i) The Cartan subalgebras of A° are exactly the centralizers of the radical complements of A.

(ii) All Cartan subalgebras of A° are conjugated by the normal subgroup $rad(A)$ of the star group of A.

Proof. ad(i): Let T be a radical complement of A. By using [74] (corollary 2.2.3, remarks 3.2.17 and 2.33) the algebra (K, A) fulfills the preconditions of theorem 24, and (K, T) is a radical complement of $rad((K, A)) = (K, rad(A))$. Thus theorem 24 implies that $C_{(K,A)}((K,T))$ is a Cartan subalgebra of $(K, A)^\circ$. As a consequence of proposition 13 the algebra $C_A(T)$ is a Cartan subalgebra of A°.

Let H be a Cartan subalgebra of A°. By using proposition 8 the set H is a subalgebra of A, and hence using proposition 13 the algebra (K, H) is Cartan subalgebra of $(K, A)^\circ$. Theorem 24 lets us deduce that a radical complement T of A exists such that $(K, H) = C_{(K,H)}(T)$ is valid. A possesses a radical complement S in A (see theorem 2.2.4 in [74]), and by using remark 2.3.3 in [74] the algebra (K, S) is a radical complement in (K, A). By using the theorem of Wedderburn-Malcev an element $u \in rad((K, A))$ exists such that $T = (K, S)^{1+u}$ is valid. Corollary 2.2.3 in [74] implies the existence of an element $r \in rad(A)$ such that $1 + u = (1, r)$ is valid. Proposition 13 is used to prove $(K, H) = C_{(K,A)}((K,S))^{(1;r)} = (K, C_A(S))^{(1;r)} = (K, C_A(S^{(r)}))$, and hence $H = C_A(S^{(r)})$ is valid. Corollary 2.3.7 in [74] implies that S and $S^{(r)}$ are radical complements of A.

ad(ii): This statement is a consequence of part(i), proposition 13 and corollary 2.3.7 in [74].⋄

Section 5.5

Separable subalgebras

$\mathcal{J}A$ — maximal torus of A
↓
reduced complements
all conjugated under
$1 + \mathcal{J}(A)$ wrt $\mathcal{J}(A)$ ⚹

$C_A(\cdot)$ ↘

A°

e_{A° — Cartan subalgebra of A°
↑
central, tors of reduced
complements
all conjugated under
$1 + \mathcal{J}(A)$ wrt $\mathcal{J}(A)$ ⚹

$C_A(T) \in e_{A^\circ}$

Diagram:
- $\mathcal{J}$
- $\mathcal{J}(A)$, $\mathcal{J}(A/T)$
- $C_A(T) \in e_{A^\circ}$
- $T \in \mathcal{J}A$ (separable + commutative)
- A°
- arrow labeled $\mathcal{N}\mathcal{S}E\mathcal{J}(\cdot)$

Special case 2

$T = \langle e_1, ..., e_n \rangle_k$ pw. orth. idempotents
$\cong k^n \sim$ splits

$C_A(T) = \bigoplus_{i=1}^n e_i A e_i$ Pierce
                     decomposition

Special case 1

$T = C_A(T)$
     ⇓
$\mathcal{J}A = e_{A^\circ}$ ~ F_q Brauer
                  Jacobson, Tits
                  ~ upper and lower
                  triangular matrices

5.5.3 Solvable group algebras

Let K be a field and G a finite group. By using 3.2.20 in [74] the group algebra KG is solvable if and only if G is abelian or $char(K) = p$ is valid and G' – the derived subgroup of G – is a p-group. If G is abelian, then $(KG)^\circ$ is nilpotent. Let $char(K) = p$ and G' be a p-group. As G' is a normal p-subgroup of G it is straightforward to prove that G possesses (with respect to Sylow's theorem) exactly one (normal) p-Sylow subgroup P. The theorem of Schur-Zassenhaus[4] implies the existence of an abelian complement H of P in G. If

---

[4] Hans Julius Zassenhaus (born 28. May 1912 in Koblenz, died 21. November 1991 in Columbus, Ohio) was a German mathematician, known for work in many parts of abstract algebra, and as a pioneer of computer algebra. He was born in Koblenz in 1912. His father was a historian and advocate for Reverence for Life as expressed by Albert Schweitzer. Hans had two brothers, Guenther and Wilfred, and sister Hiltgunt, who wrote an autobiography in 1974. According to her, their father lost his position as school principal due to his philosophy. She wrote: Hans, my eldest brother, studied mathematics. My brothers Guenther and Wilfred were in medical school. Only students who participated in Nazi activities would get scholarships. That left us out. Together we made an all-out effort. Soon our house became a beehive. Day in and day out for the next four years a small army of children of all ages would arrive to be tutored. At the University of Hamburg Zassenhaus came under the influence of Emil Artin. As he wrote later: His introductory course in analysis that I attended at the age of 17 converted me from a theoretical physicist to a mathematician. When just 21, Zassenhaus was studying composition series in group theory. He proved his butterfly lemma that provides a refinement of two normal chains to isomorphic central chains. Inspired by Artin, Zassenhaus wrote a textbook Lehrbuch der Gruppentheorie that was later translated as Theory of Groups. His thesis was on doubly transitive permutation groups with Frobenius groups as stabilizers. These groups are now called Zassenhaus groups. They have had a deep impact on the classification of finite simple groups. He obtained his doctorate in June 1934, and took the teachers' exam the next May. He became a scientific assistant at University of Rostock. In 1936 he became assistant to Artin back in Hamburg, but Artin departed for the USA the following year. Zassenhaus gave his Habilitation in 1938. According to his sister Hiltgunt, Hans was called up as a research scientist at a weather station: 55 for his part in the German war effort. Zassenhaus married Lieselotte Lohmann in 1942. The couple raised three children: Michael (born 1943), Angela (born 1947), and Peter (born 1949). In 1943 Zassenhaus became extraordinary professor. He became Managing Director of the Hamburg Mathematical Seminar. After the war, and as a fellow of the British Council, Zassenhaus visited University of Glasgow in 1948. There he was given an honorary Master of Arts degree. The following year he joined the faculty of McGill University where the endowments of Peter Redpath financed a professorship. He was at McGill for a decade with leaves of absence to Institute for Advanced Study (55/6) and California Institute of Technology (58/9). There he was using computers to advance number theory. In 1959 Zassenhaus began teaching at University of Notre Dame and became director of its computing center in 1964. Zassenhaus was a Mershon visiting professor at Ohio State University in the fall of 1963. In 1965 he came to Ohio State permanently. The mathematics department was led by Arnold Ross; Zassenhaus found a home there until his retirement in 1982. Nonetheless, he continued to take leaves of absence for visits to Göttingen (summer 67), Heidelberg (summer 69), UCLA (fall 70), Warwick (fall 72), CIT (74/5), U Montreal (77/8), Saarbrücken (79/80). He served as editor in chief of the Journal of Number Theory from its first issue in 1967. He won a Lester R. Ford Award in 1968. Hans Zassenhaus died in Columbus, Ohio on November 21, 1991. His doctoral students include Joachim Lambek.

α is the linearization of the canonical epimorphism from G onto the factor group G/P, then the identity $Kern\,\alpha = KG\,Aug(KP) = Aug(KP)KG$ is well-known ($Aug(KP)$ the augmentation ideal of KP). By using a theorem of Wallace $Aug(KP)$ is nilpotent, and thus $Kern\,\alpha$ is nilpotent, too. The factor algebra by KG modulo $Kern\,\alpha$ is isomorphic to $K(G/P)$ and thus to KH. KH is by Maschke's theorem semisimple and hence separable (see 1.9.4 in [74]). We deduce $rad(KG) = KG\,Aug(KP)$, and KH is a separable and commutative radical complement in KG.

By theorem 24 we deduce that the Cartan subalgebras of $(KG)^\circ$ are exactly the conjugates under $1 + rad(KG)$ of $C_{KG}(KH) = C_{rad(KG)}(KH) \oplus KH$. (Here we use that H is abelian because of $G' \subseteq P$.)

Its straightforward to prove that the set $\{(a-1)h \mid 1 \neq a \in P, h \in H\}$ is a K-basis of $rad(KG)$, and this result can be used to determine $C_{rad(KG)}(KH)$. The centralizer of KH in KG can be calculated as follows: H acts by conjugation on G, and thus G decomposes into H-orbits $B_1, \ldots, B_l$. The set $\{\overline{B_i} \mid i \in \underline{l}\}$ is a K-basis of $C_{KG}(KH)$.⋄

5.5.4 Solvable group algebras of dihedral groups

(i) Let G be a group, $n \in \mathbb{N}$ and $a, b \in G$, such that $o(a) = n$, $o(b) = 2$, $G = \langle a, b \rangle$ and $a^b = a^{-1}$ are valid. Then G is a dihedral group of order $2n$ symbolized by D_{2n}.

Because of $a^{-1}b^{-1}ab = a^{-1}a^b = a^{-2}$ we deduce $\langle a^2 \rangle \leq G' \leq \langle a \rangle$. If 2 is no divisor of n, then $o(a^2) = o(a)$ and $G' = \langle a \rangle$ are valid. In the other case $o(a^2) = \frac{n}{2}$ is valid, and by using $(a^2)^b = (a^b)^2 = a^{-2}$ the subgroup $\langle a^2 \rangle$ is a normal one of G with abelian factor group. Hence we deduce $G' = \langle a^2 \rangle$.

(ii) In addition, let K be a field. By using 3.2.20 in [74] the group algebra KG is solvable if and only if G is abelian or $char(K) = p$ is valid and G' is a p-group. With respect to (i) the following cases are to be analyzed (p a prime number different from 2):

(a) G is abelian.
(b) G is a 2-group and $char(K) = 2$ is valid.
(c) n is a power of p and $char(K) = p$ is valid.
(d) $\frac{n}{2}$ is a power of p and $char(K) = p$ is valid.

We will describe the Cartan subalgebras of $(KG)^\circ$ for all these cases. For this we will use the results of section 5.5.3 frequently.

(iii)(a) Let G be abelian. This is true only for $n \in \underline{2}$ true, and in this case KG is Lie nilpotent.

(iii)(b) Let G be a 2-group and K a field with $char(K) = 2$. By using a theorem of Wallace we derive that the augmentation ideal is nilpotent and $KG = Aug(KG) \oplus K1_G$ is valid. The associative nilpotency of $Aug(KG)$ implies the nilpotency of $Aug(KG)^\circ$. As $K1_G$ is central, $(KG)^\circ$ is nilpotent.

(iii)(c) Let n be a power of p and $char(K) = p$. $G' = \langle a \rangle$ is the p-Sylow subgroup of G with complement $\langle b \rangle$. $K\langle b \rangle$ is a radical complement in KG with K-basis $\{1, b\}$, and the set $\{a^s - 1, (a^s - 1)b \mid s \in \underline{n-1}\}$ is a K-basis of the nilradical. The Cartan subalgebras of $(KG)^\circ$ are the conjugates under $1 + rad(KG)$ of $C_{KG}(K\langle b \rangle) = C_{rad(KG)}(K\langle b \rangle) \oplus K\langle b \rangle$. We calculate the centralizer of $K\langle b \rangle$ in $rad(KG)$ and prove that it possesses the dimension $(n-1)$. Hence all Cartan subalgebras of $(KG)^\circ$ are $(n+1)$-dimensional.

Let $x \in rad(KG)$, like $x = \sum_{i=1}^{n-1} k_i(a^i - 1) + \sum_{j=1}^{n-1} l_j(a^j - 1)b$. We calculate:

$$\forall y \in K\langle b \rangle : xy = yx \iff$$

$$\sum_{i=1}^{n-1} k_i a^i b + \sum_{j=1}^{n-1} l_j a^j - \sum_{i=1}^{n-1} k_i b a^i - \sum_{j=1}^{n-1} l_j b a^j b = 0 \iff$$

$$\sum_{i=1}^{n-1} k_i(a^i b - a^{-i} b) + \sum_{j=1}^{n-1} l_j(a^j - a^{-j}) = 0 \iff$$

$$\sum_{i=1}^{n-1} k_i(a^i b - a^{n-i} b) + \sum_{j=1}^{n-1} l_j(a^j - a^{n-j}) = 0 \iff$$

$$\sum_{i=1}^{\frac{n-1}{2}} (k_i - k_{n-i}) a^i b + \sum_{j=1}^{\frac{n-1}{2}} (l_j - l_{n-j}) a^j = 0 \iff$$

$$\forall i, j \in \underline{\frac{n-1}{2}} : k_i = k_{n-i} \land l_j = l_{n-j} \quad .$$

(iii)(d) Let $\frac{n}{2}$ be a power of p, like $n = 2p^r$, and $char(K) = p$. $G' = \langle a^2 \rangle$ is the p-Sylow subgroup of G with complement $H := \{1, b, a^{p^r}, a^{p^r} b\}$. KH is a radical complement in KG with K-basis H, and the set $\{a^{2s} - 1, b(a^{2s} - 1), a^{p^r} a^{2s} - 1, a^{p^r} b a^{2s} - 1 \mid s \in \underline{p^r - 1}\}$ is a K-basis of the nilradical. The Cartan subalgebras are the conjugates under $1 + rad(KG)$ of $C_{KG}(KH) = C_{rad(KG)}(KH) \oplus KH$. We calculate the centralizer of KH in $rad(KG)$ and prove that it is $(n-2)$-dimensional. Hence all Cartan subalgebras of $(KG)^\circ$ are of dimension $n + 2$.

Let $x \in rad(KG)$, like

$$x = \sum_{i=1}^{p^r-1} l_i(a^{2i}-1) + \sum_{i=1}^{p^r-1} m_i b(a^{2i}-1)$$
$$+ \sum_{i=1}^{p^r-1} r_i a^{p^r}(a^{2i}-1) + \sum_{i=1}^{p^r-1} s_i a^{p^r} b(a^{2i}-1).$$

x centralizes KH if and only if $x \circ b = 0 = x \circ a^{p^r}$ is valid.
We calculate:

$$x \circ a^{p^r} = \sum_{i=1}^{p^r-1} m_i(ba^{2i}a^{p^r} - a^{p^r}ba^{2i} - ba^{p^r} + a^{p^r}b)$$
$$+ \sum_{i=1}^{p^r-1} s_i(a^{p^r}ba^{2i}a^{p^r} - ba^{p^r} - a^{p^r}ba^{p^r} + b)$$
$$= 0,$$

because a^{p^r} is an involution.
In addition, the following calculation is valid:

$$x \circ b = \sum_{i=1}^{p^r-1} l_i(a^{2i}b - ba^{2i}) + \sum_{i=1}^{p^r-1} m_i(ba^{2i}b - a^{2i})$$
$$+ \sum_{i=1}^{p^r-1} r_i(a^{p^r}a^{2i}b - ba^{p^r}a^{2i} - a^{p^r}b + ba^{p^r})$$
$$+ \sum_{i=1}^{p^r-1} s_i(a^{p^r}ba^{2i}b - ba^{p^r}ba^{2i} - a^{p^r}bb + ba^{p^r}b)$$
$$= \sum_{i=1}^{p^r-1} l_i(a^{2i} - a^{-2i})b + \sum_{i=1}^{p^r-1} (-m_i)(a^{2i} - a^{-2i})$$
$$+ \sum_{i=1}^{p^r-1} r_i(a^{2i} - a^{-2i})a^{p^r}b + \sum_{i=1}^{p^r-1} (-s_i)(a^{2i} - a^{-2i})a^{p^r}$$
$$= \sum_{i=1}^{\frac{p^r-1}{2}} (l_i - l_{p^r-i})a^{2i}b + \sum_{i=1}^{\frac{p^r-1}{2}} (m_{p^r-i} - m_i)a^{2i}$$
$$+ \sum_{i=1}^{\frac{p^r-1}{2}} (r_i - r_{p^r-i})a^{2i}a^{p^r}b + \sum_{i=1}^{\frac{p^r-1}{2}} (s_{p^r-i} - s_i)a^{2i}a^{p^r}.$$

Thus x centralizes b if and only if for all $i \in \frac{p^r-1}{2}$, the conditions $l_i = l_{p^r-i}$, $m_i = m_{r-i}$, $r_i = r_{p^r-i}$ and $s_i = s_{p^r-i}$ are valid.⋄

5.5.5 Solvable group algebras of quaternion groups

(i) Let G be a group, $n \in \mathbb{N}_{\geq 2}$ and $a, b \in G$ such that $o(a) = 2n$, $o(b) = 4$, $G = \langle a, b \rangle$, $a^b = a^{-1}z$ and $z = a^n = b^2$ are valid. Then G is called a quaternion group of order $4n$ – noted by Q_{4n}. It is straightforward to prove that the derived subgroup is cyclic and of order n ist generated by a^2. Hence its factor group is of order 4.

(ii) In addition, let K be a field. By using 3.2.20 in [74] the group algebra KG of the group algebra KG is solvable if and only if G is abelian or $char(K) = p$ is valid and G' is a p-group. With respect to (i) the following cases are to be analyzed (p a prime number different from 2):

(a) G is abelian.
(b) G is a 2-group, $char(K) = 2$
(c) n is a power of p, $char(K) = p$.

We will describe the Cartan subalgebras of $(KG)°$ for all these cases. For this we will use the results of section 5.5.3 frequently.

(iii)(a) Let G be abelian. This case is not possible.

(iii)(b) Let G be a 2-group and $char(K) = 2$. By using a theorem of Wallace we derive that the augmentation ideal is nilpotent, and $KG = Aug(KG) \oplus K1_G$ is valid. The associative nilpotency of $Aug(KG)$ implies the nilpotency of $Aug(KG)°$. As $K1_G$ is central, and thus $(KG)°$ is nilpotent.

(iii)(c) Let n be a power of p and $char(K) = p$. $G' = \langle a^2 \rangle$ is the p-Sylow subgroup of G with complement $\langle b \rangle$. $K\langle b \rangle$ is a radical complement in KG with K-Basis $\{1, b, b^2, b^3\}$, and the set $\{a^s - 1, (a^s - 1)b, (a^s - 1)b^2, (a^s - 1)b^3 \mid s \in \underline{n}\}$ is a K-basis for the nilradical of KG. The Cartan subalgebras are the conjugates under $1 + rad(KG)$ of $C_{KG}(K\langle b \rangle) = C_{rad(KG)}(K\langle b \rangle) \oplus K\langle b \rangle$. We determine this centralizer of $K\langle b \rangle$ in $rad(KG)$ and prove that its $2(n-1)$-dimensional. Hence all Cartan subalgebras of $(KG)°$ are of dimension $2n+2$. Let $z \in rad(KG)$, like

$$z = \sum_{i=1}^{n} k_i(a^{2i} - 1) + \sum_{i=1}^{n} l_i(a^{2i} - 1)b + \sum_{i=1}^{n} m_i(a^{2i} - 1)b^2 + \sum_{i=1}^{n} r_i(a^{2i} - 1)b^3.$$

z commutes with all elements of $K\langle b\rangle$ if and only if it commutes with b. This is equivalent to

$$\sum_{i=1}^{n} k_i(a^{2i} - a^{-2i})b + \sum_{i=1}^{n} l_i(a^{2i} - a^{-2i})b^2 +$$
$$\sum_{i=1}^{n} m_i(a^{2i} - a^{-2i})b^3 + \sum_{i=1}^{n} r_i(a^{2i} - a^{-2i})1 = 0.$$

Hence the centralizer is of dimension

$$4 \cdot \tfrac{n-1}{2} = 2(n-1).\diamond$$

5.5.6 Splitting solvable algebras

In this section we specialize our analysis to splitting solvable algebras. These are associative algebras for which the factor algebra by its nilradical is isomorphic to a n-fold direct product of the base field. In particular, those algebras are solvable and its factor algebra by the nilradical is separable. For example, every associative solvable $\mathbb{C}$-algebra is splittable, the Solomon-Tits algebras are splittable over every base field as well as the Solomon algebras in characteristic zero and the algebras of upper and lower triangular matrices.

Lemma 5 *Let K be a field, A an associative finite-dimensional K-algebra and let $e_1, \cdots, e_n$ be pairwise orthogonal idempotents of A such $T := \langle e_1, \cdots, e_n \rangle_K$ is a complement of $rad(A)$ in A. The following statements are valid:*

(i) *K is a splitting field of A. In particular, A is solvable and $A/rad(A)$ is separable.*

(ii) *$\sum_{i=1}^{n} e_i$ is the identity element of T. In particular, T is unitary.*

(iii) *If A is unitary, then $1_A = 1_T$ is valid. In particular, T is unital.*

(iv) *$T \subseteq C_A(T)$*

(v) *If A is unitary, then $C_A(T) = \bigoplus_{i=1}^{n} e_i A e_i$ is valid.*

(vi) *If A is unitary, then T is self-centralizing in A if and only if for all $i \in \underline{n}$ the Pierce component $e_i A e_i$ is contained in T.*

(vii) *If A is unitary and T self-centralizing in A, then the following statements are valid:*

(a) $A = \bigoplus_{i,j=1}^{n} e_i A e_j$

(b) $rad(A) = \bigoplus_{i \neq j=1}^{n} e_i A e_j$

(c) $T = \bigoplus_{i=1}^{n} e_i A e_i$

(d) $\forall i \in \underline{n} : e_i A e_i = \langle e_i \rangle_K$

(viii) If A is unitary, then T is self-centralizing in A if and only if for all $i \in \underline{n}$ the Pierce component $e_i A e_i$ is one-dimensional (and thus identical to $\langle e_i \rangle_K$).

(ix) If A is unitary, then T is self-centralizing in A if and only if $T = \bigoplus_{i=1}^{n} e_i A e_i$ is valid.

Proof. ad(i)+(ii): It is well-known that T is isomorphic to K^n and that $\sum_{i=1}^{n} e_i$ as an identity of T.

ad(iii): This part is a direct consequence of remark 1.10.1 in [74].

ad(iv): This statement is a consequence of the commutativity of T (see part (i)).

ad(v): $e_1, \cdots, e_n$ are idempotent and pairwise orthogonal. Hence $\bigoplus_{i=1}^{n} e_i A e_i \leq C_A(T)$ is valid. Let $a \in C_A(T)$. By using part (iii) we derive $A = \bigoplus_{i,j=1}^{n} e_i A e_j$, and, in addition, we take for all $i, j \in \underline{n}$ elements $a_{i,j} \in A$, such that $a = \sum_{i,j=1}^{n} e_i a_{i,j} e_j$ is valid. Let $r \in \underline{n}$. Then $ae_r = \sum_{i=1}^{n} e_i a_{i,r} e_r$ and $e_r a = \sum_{i=1}^{n} e_r a_{i,r} e_i$ are valid. By using $a \circ e_r = 0$ and $A = \bigoplus_{i,j=1}^{n} e_i A e_j$ we deduce $e_i a_{i,r} e_r = 0 = e_r a_{r,i} e_i$. Thus part (v) is proven.

ad(vi): This part is a direct consequence of parts (iv) and (v).

ad(vii): Part (a) is deductable by the two-sided Pierce decomposition and part (ii), and part (b) is deducible by part (v). By using a dimension argument we derive part (d) from by part (c). For all $i \neq j \in \underline{n}$ the identity $(e_i A e_j)(e_i A e_j) = 0$ is valid. Hence the Pierce component $e_i A e_j$ is nilpotent. A is solvable (see part (i)). Thus $e_i A e_j$ – using proposition 5 in [76] – is contained in the nilradical of A. The proof of this part is now a consequence of parts (a) and (c) as well as by a dimension argument.

ad(viii): This part is a direct consequence of parts (v) and (vi).

ad(ix): This part is a direct consequence of parts (v) and (vi).⋄

A consequence of this lemma and of theorems 23 and 24 is:

Theorem 26 *Let K be a field, A an associative finite-dimensional K-algebra and let $e_1, \cdots, e_n$ be pairwise orthogonal idempotents of A such $T := \langle e_1, \cdots, e_n \rangle_K$ is a complement of $rad(A)$ in A. The following statements are valid:*

(i) The maximal tori of A° are exactly the conjugates of T under $1 + rad(A)$.

(ii) The Cartan subalgebras of A° are exactly the conjugates of $C_A(T) = \bigoplus_{i=1}^{n} e_i A e_i$ under $1 + rad(A)$.⋄

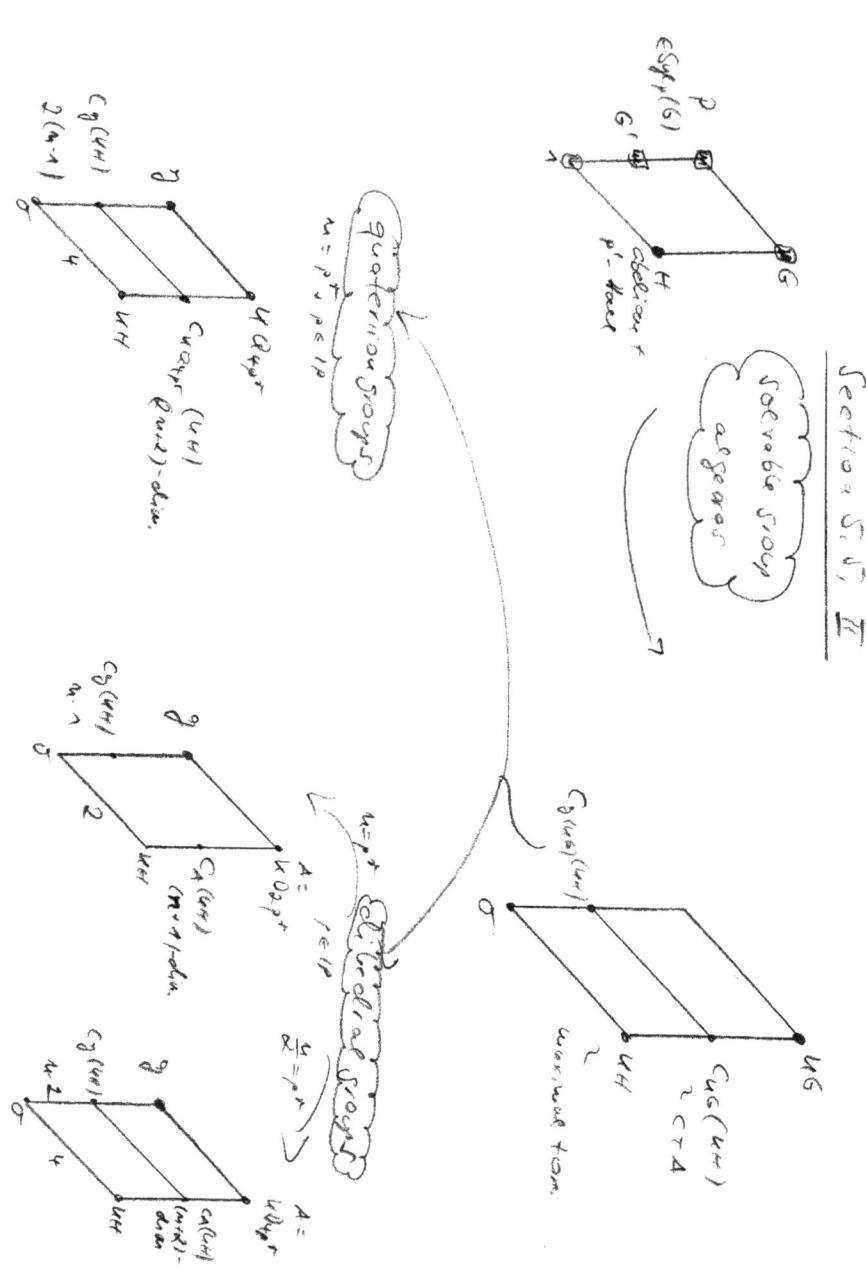

5.5.7 Open-ended questions and exercises

Open-ended questions 5 *Is there a description of the maximal tori for a solvable algebra without using the condition of separability of the radical factor algebra?*

Excercise 122 *With respect to theorems 25 and 24 analyze the following statement: The maximal tori of A coincide with the Cartan subalgebras of A° if and only if a self-centralizing radical complement exists.*

Excercise 123 *Let $n \in \mathbb{N}$ and K be a field. On what terms is KS_n a solvable algebra?*

Excercise 124 *Let $n \in \mathbb{N}$ and K be a field. On what terms is $K^{n \times n}$ solvable?*

Excercise 125 *Let $n \in \mathbb{N}$ and K be a field. The subalgebras of upper and lower triangular matrices of $K^{n \times n}$ are solvable and unital. What is their dimension? Are they isomorphic? Are they anti-isomorphic? What is their associative nilradical? Do radical complements exist? Is it possible to describe all radical complements? (Tip: see [74])*

Excercise 126 *Let $n \in \mathbb{N}$ and K be a field. Determine all maximal tori and Cartan subalgebras of the Lie algebra associated to the algebra of lower triangular matrices of $K^{n \times n}$!*

Excercise 127 *Let A, B be associative unitary K-algebras. In what way are the maximal tori and Cartan subalgebras of A° connected to those of B° with respect to an isomorphism or anti-isomorphism between A and B?*

Excercise 128 *Let $n \in \mathbb{N}$ and K be a field. Determine all maximal tori and Cartan subalgebras of the Lie algebra associated to the algebra of upper triangular matrices of $K^{n \times n}$!*

Excercise 129 *Prove proposition 13!*

Excercise 130 *Prove definition and remark 3!*

Excercise 131 *Prove definition and remark 2!*

Excercise 132 *Let K be a field, G a finite group and N a normal subgroup of G. The natural epimorphism from G onto G/N can be extended to a K-linear epimorphism from KG onto $K(G/N)$. This extension is an epimorphism of algebras with kernel $KGAug(KN) = Aug(KN)KG$. Calculate a basis and the dimension of this kernel.*

Excercise 133 Let A be an associative (unitary) algebra and r a nilpotent element of A. The element $1 + r$ is invertible with respect to $\cdot$ and r is invertible with respect to $\star$. The formula $(1+r)^{-1} = 1 + r^{(-1)}$ is valid. In addition, prove that $1 + rad(A)$ resp. $rad(A)$ is a normal subgroup with respect to $\cdot$ resp. $\star$.

Excercise 134 Prove the statements of part (iii) in example 1! Is there a connection between the maximal tori and the Cartan subalgebras of a Lie algebra of an associative algebra A and the algebra connected with A by adjoining an identity element?

Excercise 135 In case of a solvable group algebra KG and a subgroup H of G decompose the orbit-sums related to the action of H by conjugation on G in a sum of a nilpotent and a fully-separable part (Both summands do not need to commute.).

Excercise 136 (eAe) Let K be a field, A an associative finite-dimensional unitary K-algebra with a separable factor algebra by its nilradical, T a commutative radical complement of $rad(A)$ in A and e an idempotent of A. Prove the following statements:

(i) e is diagonalizable and hence fully-separable.

(ii) The subalgebra generated by $\{1, e\}$ is separable. By using the extended version of the theorem of Wedderburn-Malcev there is a radical complement conjugated to T by $1 + r$ in which $r \in rad(A)$ is valid.

(iii) Do a search in the literature to find the result that $e\,rad(A)\,e$ is the nilradical of eAe.

(iv) By using $A = rad(A) \oplus T^{1+r}$ deduce the formula $eAe = e\,rad(A)\,e \oplus eT^{1+r}e$. Prove that $eT^{1+r}e$ is a separable radical complement. Hence eAe a finite-dimensional associative solvable unitary K-Algebra with separable factor algebra by its radical. What is the unit element of eAe?

(v) Deduce by the previous parts that the maximal tori of A are the conjugates of T by $1 + rad(A)$.

(vi) Deduce by the previous parts that the maximal tori of eAe are the conjugates of $eT^{1+r}e$ by $e + e\,rad(A)\,e$.

(vii) By using centralizers describe the Cartan subalgebras of A° and $(eAe)^\circ$.

(viii) In case of a central idempotent e deduce the formula $C_{eAe}(eT^{1+r}e) = eC_A(T^{1+r})e$.

(ix) What is the nilradical of $(eAe)^\circ$?

Excercise 137 *(zero-extension)* We focus again on exercise 39 about zero-extensions over a field K. On what terms is this algebra solvable? On what terms is the factor algebra by the nilradical of the zero-extension isomorphic to K^n for a natural number $n \in \mathbb{N}$? For both cases determine the maximal tori and the Cartan subalgebras! Is there a connection to the original algebra used for the construction of the zero-algebra? On what terms are the maximal tori and Cartan subalgebras self-centralizing?

Excercise 138 Let K be a field, $n \in \mathbb{N}$, A an associative K-algebra and C a Cartan subalgebra of A°. The diagonal matrices in $A^{n \times n}$ with entries in C are a Cartan subalgebra of $(A^{n \times n})^\circ$. In addition, analyze under what terms this Cartan subalgebra is self-centralizing.

5.6 Lie nilpotent associative algebras

Proposition 8 lets us deduce that Cartan subalgebras in Lie algebras associated to associative algebras are associative subalgebras. Therefor we analyze in this section the associative structure of these subalgebras. In particular, we clarify on what terms an associative algebra is Lie nilpotent. The result is intimately connected to the question on what terms the group of units is nilpotent. We demonstrate the results of this section by the group algebra.

5.6.1 Lie nilpotency

A first characterization of Lie nilpotent associative algebras is a direct consequence of our main theorem 1:

Corollary 1 *Let A be a finite-dimensional associative unitary K-algebra. A° is nilpotent if and only if every fully separable element of A is central.*

Proof. By using our main theorem 1 we deduce that A° is nilpotent if every maximal torus of A is central. If a is a fully separable element of a, then $K[a]$ is a torus (see lemma 5.2.5 in [74]) which is contained in a maximal torus of A.⋄

We generalize the corollary to the following theorem:

Theorem 27 *Let A be a finite-dimensional associative unitary K-algebra. The following statements are equivalent:*

(i) A° is nilpotent.

(ii) Every fully separable element of A is central.

(iii) A possesses exactly one maximal separable subalgebra. This subalgebra is central.

(iv) The set of fully separable elements is a central subalgebra.

Proof. By using corollary 1 the parts (i) and (ii) are equivalent. With respect to theorems in [74] the set of fully separable elements is a subalgebra in commutative finite-dimensional associative algebras. Hence (iv) is a consequence of (ii). Because the elements of a separable commutative algebra are fully separable (see again results in [74]) part (iii) can be deduced by (iv). If t is a fully separable element, then the subalgebra generated by t is a separable subalgebra (see again results in [74]), and part (iii) is proven.⋄

Remark 10 (i) Let A be a finite-dimensional associative K-algebra with central radical complement. For all $n \in \mathbb{N}$ the identity

$$\underbrace{A \circ \cdots \circ A}_{n-mal} = \underbrace{rad(A) \circ \cdots \circ rad(A)}_{n-mal} \leq rad(A)^n$$

is valid. By using the associative nilpotency of $rad(A)$ we deduce the Lie nilpotency of $rad(A)^{\circ}$. Hence A° is nilpotent with $cl(A^{\circ}) \leq cl(rad(A))$.

(ii) We focus on the algebra within section 5.4.4. By the multiplication table we deduce $i \circ j = i$. Hence for all $n \in \mathbb{N}$ the identity $i\,ad(j)^n = i$ is valid, and therefor the quaternion algebra in characteristics two is not Lie nilpotent.⋄

A more detailed description of the associative structure of Lie nilpotent associative algebra is contained within the following theorem. We remark that for a non-unitary associative K-algebra an element is fully separable over K if its fully separable over K as element of the algebra constructed by the adjunction of a unit.

Theorem 28 *Let A be a finite-dimensional associative K-algebra with separable factor algebra by its nilradical. The following statements are equivalent:*

(i) A° is nilpotent.

(ii) A possesses exactly one central radical complement.

(iii) A is solvable and possesses exactly one radical complement.

(iv) A is solvable and the set of fully separable elements is a radical complement.

(v) A is solvable and the set of fully separable elements is a K-subspace.

Proof. (ii) → (i): see remark 10

(i) → (ii): Let A° be nilpotent. A is with respect to lemma 2 solvable. Let T be a commutative radical complement (see 2.2.4 in [74]) of A. By using theorem 25 the identity $C_A(T) = A$ is valid, and hence T is central in A. The proof of (ii) is finished by using corollary 2.3.7 in [74].

(ii) → (iii): The radical complement is central, and hence it is commutative and therefor A solvable. In addition, A possesses – using 2.3.7 in [74] – exactly one radical complement.

(iii) → (ii): Let A be solvable and possessing exactly one radical complement. By using corollary 5.1.5 in [74] the intersection of all radical complements is central.

(iv) → (iii): The set of all fully separable elements of A is invariant under conjugation with the quasiregular group of A (see 2.3.6 in [74]). Hence (iii) is a consequence of (iv).

(iii) → (iv): Let T be the radical complement of A and A be solvable. Then T is a commutative and separable subalgebra of A. By using theorem 5.6.11 in [74] every element of T is fully separable over K. If a is a fully separable element of A, then $K[a]$ is with respect to the same theorem 5.6.11 a separable subalgebra of A. Corollary 2.3.7 in [74] implies that this subalgebra is contained in a radical complement of A. T is the only radical complement, and hence (iv) is proven.

(iv) → (v): This statement is straightforward to prove.

(v) → (iv): Let T be a radical complement of A. Then T is commutative and separable. By using theorem 5.6.11 in [74] we deduce that every element of T is fully separable over K. The set of fully separable elements is direct to the nilradical because the only element which is nilpotent and fully separable is the zero element. Comparing dimensions yields to the statement that T is exactly the set of fully separable elements of A.◇

The following corollary generalizes remark 8 and will be useful to determine a Cartan subalgebra of the tensor product later on:

Corollary 2 *Let A, B be finite-dimensional associative K-algebras with separable factor algebras by their nilradicals. If A° and B° are nilpotent, then $(A \otimes B)^\circ$ is nilpotent, too, and $A \otimes B$ possesses a separable factor algebra by its nilradical.*

Proof. Let A, B be as Lie algebras nilpotent. If T resp. S is a radical complement of A resp. B, then these complements are central by using theorem 28. With respect to theorem 2.2.9 in [74] we deduce that $T \otimes S$ is a separable radical complement of $A \otimes B$. $Z(A \otimes B) = Z(A) \otimes Z(B)$ is valid and hence $T \otimes S$ is central, too, and therefor $A \otimes B$ is – again by using theorem 28 – Lie nilpotent.⋄

5.6.2 Nilpotent group of units

Proposition 14 *Let A be a finite-dimensional associative unitary K-algebra.*

(i) *If $rad(A)$ possesses a central radical complement, then $E(A)$ is nilpotent.*

(ii) *If T is a radical complement of A, then $1_A \in T$ is valid.*

(iii) *If T is a radical complement of A and $E(T)$ is central in A, then $E(A)$ is the inner direct product of the nilpotent normal subgroup $1_A + rad(A)$ and the central normal subgroup $E(T)$. In particular, $E(A)$ is nilpotent.*

Proof. ad(i): see 1.1.8 in [75]

ad(ii): see 1.10.1 in [74]

ad(iii): see 1.1.8 in [75].⋄

Within the next lemma the theorem of Scott presented in chapter 3 is used:

Lemma 6 *Let A be a finite-dimensional associative unitary K-algebra. If $E(A)$ is nilpotent, then A is solvable.*

Proof. By using lemma A.1.1 in [74] the identity $E(A/rad(A)) = E(A)/(1+rad(A))$ is valid, and the nilpotency of $E(A)$ is passed to its factor group. Let $D_1, \cdots, D_r$ be K-division algebras and $n_1, \cdots, n_r \in \mathbb{N}$, such that $A/rad(A)$ is isomorphic to $D_1^{n_1 \times n_1} \times \cdots \times D_r^{n_1 \times n_r}$. By induction and using 1.1.8 in [75] we derive that $E(A)/(1 + rad(A))$ is isomorphic to $E(D_1^{n_1 \times n_1}) \times \cdots \times E(D_r^{n_1 \times n_r})$. In particular, for every $i \in \underline{r}$ the group $E(D_i)$ is nilpotent. The theorem 8 of Scott implies that for every $i \in \underline{r}$ the skew field D_i is a field. In addition, for all $i \in \underline{r}$ the groups $GL(n_i, K)$ are nilpotent by our assumption. Besides $GL(2,2)$ and $GL(2,3)$ these groups are for $n_i \geq 2$ by using results on page 181 in [25] non-solvable. On page 183 in [25] it is proven that $GL(2,2)$ resp. $PSL(2,3)$ are isomorphic to S_3 resp. A_4, and these groups are not nilpotent. Therefor we derive $n_i = 1$ for all $i \in \underline{r}$. ⋄

Theorem 29 *Let K be a field with at least three elements and A an associative finite-dimensional unitary K-algebra with separable factor algebra by its nilradical. The following statements are equivalent:*

(i) $E(A)$ *is nilpotent.*

(ii) A *possesses a central radical complement.*

(iii) A° *is nilpotent.*

Proof. (ii) $\leftrightarrow$ (iii): see theorem 28

(ii) $\to$ (i): see proposition 14

(i) $\to$ (ii): Let $E(A)$ be nilpotent. Then A is – by using lemma 6 – solvable. Let T be a radical complement of A (Wedderburn-Malcev). Theorem 5.16 and corollary 5.18 in [4] imply that $E(C_A(T))$ is a Carter subgroup of $E(A)$. By the maximal nilpotency of the Carter subgroup we derive $E(A) = E(C_A(T))$, and, in particular, $1 + rad(A) \subseteq E(C_A(T)) \subseteq C_A(T)$ is valid. Therefor $1_A + rad(A)$ centralizes the radical complement T. The theorem of Wedderburn-Malcev lets us deduce that A possesses exactly one radical complement. Corollary 5.1.5 in [74] implies that the intersection of all radical complements of A – which is exactly T – is central.$\diamond$

Within the proof we have used a theorem of Thorsten Bauer about Carter subgroups in solvable algebras. In series II we will present this theorem in details.

Theorem 30 *Let A be an associative finite-dimensional unitary K-algebra with separable factor algebra by its nilradical. $E(A)$ is nilpotent if and only if for one (every) radical complement T of A the subgroup $E(T)$ is central. In this case $E(A)$ is the inner direct product of the nilpotent normal subgroup $1 + rad(A)$ and the central subgroup $E(T)$. If A° is nilpotent, then $E(A)$ is nilpotent, too.*

Proof. One part is included in proposition 14. Let $E(A)$ be nilpotent. Then A is solvable using lemma 6. If T is a radical complement of A, then $C_{E(A)}(E(T))$ is with respect to theorem 5.16 in [4] a Carter subgroup of $E(A)$. The maximal nilpotency implies $E(A) = C_{E(A)}(E(T))$, and thus $E(T)$ is central. Finally, the proof is finished by using 1.1.8 in [75] and theorem 28.$\diamond$

Counterexample 1 Let K be a field with two elements and $n \in \mathbb{N}_{\geq 2}$. We focus on the algebra A of lower triangular matrices over K. The Cartan subalgebras of A° are exactly the conjugates of the subalgebra of diagonal matrices under $1 + rad(A)$. In particular, A is not Lie nilpotent. The unit group of A is exactly $1 + rad(A)$ and is nilpotent.$\diamond$

By using the adjunction of a unit its possible to transfer the theorems 29 and 30 to non-unitary algebras. These proofs are left to the reader as exercises:

Theorem 31 *Let K be a field with at least three elements and A an associative finite-dimensional K-algebra with separable factor algebra by its nilradical. The following statements are equivalent:*

(i) $Q(A)$ is nilpotent.

(ii) A possesses a central radical complement.

(iii) $A°$ is nilpotent.⋄

Theorem 32 *Let A be an associative finite-dimensional K-algebra with separable factor algebra by its nilradical. $Q(A)$ is nilpotent if and only if for one radical complement T of A the group $Q(T)$ is central. In this case $Q(A)$ is the direct product of the nilpotent normal subgroup $rad(A)$ and the central normal subgroup $Q(T)$.*⋄

5.6.3 Group algebras and Lie nilpotency

Let K be a field and G a finite group. The authors of [11] and [36] have proven that KG is Lie nilpotent if and only if $E(KG)$ is nilpotent. In the case $char(K) = 0$ this is the case if and only if G is abelian. In the modular case $char(K) = p$ this is the case if and only if the subgroup G' is a p-subgroup of the nilpotent group G.

With respect to theorems 29 and 30 we analyze in the modular case on what terms KG possesses a separable factor algebra by its nilradical and a central radical complement.

For this, let $char(K) = p$ and G' a p-subgroup of the nilpotent group G. Let P be the normal p-Sylow subgroup of G with normal complement N. The condition $G' \leq P$ implies that N is central in G. If α is the linearization of the canonical group epimorphism from G onto G/P, then it is well-known that $Kern\, \alpha = KG\,Aug(KP) = Aug(KP)KG$ is valid, in which $Aug(KP)$ is the augmentation ideal of KP. By using a theorem Wallace we conclude that $Aug(KP)$ is nilpotent, and thus we derive the nilpotency of $Kern\, \alpha$, too. The factor algebra of KG modulo $Kern\, \alpha$ is isomorphic to $K(G/P)$ which is isomorphic to KN. KN is with respect to a theorem of Maschke semisimple and therefor separable by 1.9.4 in [74]. We have proven $rad(KG) = KG\,Aug(KP)$, and KN is a separable radical complement in KG. As N is central in G, the subalgebra KN is a central radical complement of KG.⋄

Section 5.6.

A° nilpotent $= C_A(T) = m_r(A^\circ)$

$\mathfrak{z}^\circ$

$C_{\mathfrak{z}}(A)$ ——— $\mathfrak{z}(A) = C_{\mathfrak{z}}(A) \oplus N^{op}(A)$

KerP(A) ~ unique maximal complement
~ set of fully separable elements

$|n| \geq 3$

A° nilpotent

$\mathfrak{z}^\circ$ ——— most unique roughly separable fully separable

$\mathfrak{z}(G)$ —— Hadamard + hess

G

$(UG)^\circ$ nilpotent

$C_g(UH)$

$\mathfrak{z}^\circ$

UH —— $\mathfrak{z}(UG)$ —— UG°

$F(A)$ nilpotent

$C_{A_g}(F(A))$

$A \rtimes g$ —— $\mathfrak{z}(F(A))$ —— $F(A)$

5.6.4 Exercises

Excercise 139 *Let K be a field, $n \in \mathbb{N}$ and A a finite-dimensional unitary associative K-algebra with separable factor algebra by its nilradical. In the following cases determine whether A°, $E(A)$ and A are solvable or nilpotent and calculate the Cartan subalgebras of A°:*

(i) $A = K\Pi_n$ the Solomon-Tits algebra (see e.g. [76])

(ii) $A = D_n$ the Solomon algebra for $char(K) = 0$ (see e.g. [4])

(iii) $A = $ The algebra of lower triangular matrices of $K^{n \times n}$

(iv) $A = $ The algebra of upper triangular matrices of $K^{n \times n}$.

Excercise 140 *With respect to theorem 27 analyze part (iii) whether the words separable, maximal and central can be permuted and the permuted version of part (iii) is still valid.*

Excercise 141 *Is a commutative and nilpotent associative algebra Lie nilpotent? On what terms is this statement true?*

Excercise 142 *Is every solvable associative algebra Lie nilpotent? On what terms is this statement true?*

Excercise 143 *Prove remark 10!*

Excercise 144 *Is the quaternion algebra in uneven characteristic Lie nilpotent? On what terms is this statement true? Solve this problem for its group of units, too.*

Excercise 145 *Is the quaternion algebra in even characteristic Lie nilpotent? On what terms is this statement true? Solve this problem for its group of units, too.*

Excercise 146 *By using corollary 2 calculate the dimensions of $A \otimes B$, of its nilradical and of all radical complements!*

Excercise 147 *If A, B are associative unitary algebras, then the identity $Z(A \otimes B) = Z(A) \otimes Z(B)$ is valid.*

Excercise 148 *Let $n \in \mathbb{N}$, K be a field and A, B the algebras of lower and upper triangular matrices of $K^{n \times n}$. By using the proof of corollary 2 determine the dimensions of $A \otimes A$, $A \otimes B$, $B \otimes A$, $B \otimes B$ and their nilradicals and factor algebras by their nilradicals!*

Excercise 149 *Prove proposition 14 in details!*

Excercise 150 Let A be a finite-dimensional associative unitary solvable K-algebra. A° and $E(A)$ are solvable, too. Is the converse statement true, too? (Tip: [74])

Excercise 151 Prove counterexample 1!

Excercise 152 Prove theorem 31!

Excercise 153 Prove theorem 32!

Excercise 154 Let A be an associative unitary K-algebra. Prove that $1 + rad(A)$ is a normal subgroup of $E(A)$ and that the identity $E(A/rad(A)) = E(A)/(1+rad(A))$ is valid. In addition, if A is right Artian, then $1+rad(A)$ is nilpotent.

Excercise 155 Let K be a field and $n \in \mathbb{N}$. On what terms are $(KS_n)^\circ$, KS_n and $E(KS_n)$ solvable or nilpotent? In the case of nilpotency determine a central radical complement!

Excercise 156 Let K be a field and $n \in \mathbb{N}$. On what terms are $(KA_n)^\circ$, KA_n and $E(KA_n)$ solvable or nilpotent? In the case of nilpotency determine a central radical complement!

Excercise 157 Let K be a field and $n \in \mathbb{N}$. On what terms are $(KD_{2n})^\circ$, KD_{2n} and $E(KD_{2n})$ solvable or nilpotent? In the case of nilpotency determine a central radical complement!

Excercise 158 Let K be a field and $n \in \mathbb{N}$. On what terms are $(KQ_{4n})^\circ$, KQ_{4n} and $E(KQ_{4n})$ solvable or nilpotent? In the case of nilpotency determine a central radical complement!

Excercise 159 Prove the following statement: On page 183 in [25] it is proven that the group $GL(2,2)$ resp. $PSL(2,3)$ is isomorphic to the group S_3 resp. A_4. These groups are not nilpotent.

Excercise 160 (zero-extension) We focus again on exercise 39. On what terms is the zero-extension nilpotent as associative algebra, nilpotent as Lie algebra and its group of units nilpotent. Analyze whether this question is closely connected to the existence of a central radical complement. If the analysis is to complex, then assume that the base field is perfect. Is the solution connected to the algebra the zero-extension is based on? Is it possible to reduce the calculation of the nilpotency class of the associated Lie algebra of the zero-extension to the algebra the zero-extension is based on?

Excercise 161 *(eAe) This exercise is closely related to exercise 136. Analyze whether eAe is as associative algebra nilpotent if and only if the algebra is Lie nilpotent resp. its group of unit is nilpotent. Analyze if this question is closely connected to the existence of a central radical complement. If the analysis is to complex, then assume that the base field is perfect and/or the idempotent is central. Is the solution connected to the algebra A? Is it possible to reduce the calculation of the nilpotency class of $(eAe)^\circ$ to the algebra A° if?*

5.7 Simple, semisimple and separable algebras

In this section we analyze Cartan subalgebras of Lie algebras associated to simple finite-dimensional associative K-algebras and we reduce the analysis for semisimple algebras to the simple ideals decomposing it.

5.7.1 Central-simple algebras

The proof of the following lemma is based on a proof by Salvatore Siciliano in [59]:

Lemma 7 *Let A be a central-simple finite-dimensional associative K-algebra. If H is a Cartan subalgebra of A°, then H is a maximal and self-centralizing torus of dimension $\mathrm{ind}(A)$ of A. In particular, H is a maximal commutative subalgebra of A.*

Proof. By using proposition 8 and lemma 2 the Lie algebra H is an unitary solvable subalgebra of A. We define $n := \mathrm{ind}(A)$. If T is a maximal subfield of A, then it is well-known that $A \otimes T$ is isomorphic to $T^{n \times n}$. If F is a algebraically closed superfield of T, then $A \otimes F \cong F^{n \times n}$ is valid, and hence we deduce $n^2 = dim_K(A) = dim_F(A \otimes F)$. By [29] the subalgebra $H \otimes F$ is a Cartan subalgebra of the F-algebra $(A \otimes F)^\circ$ which is isomorphic to $gl(n, F)$. Using again results in [29] we know that the Cartan subalgebras of $(F^{n \times n})^\circ$ are the conjugates in $GL(n, F)$ of the subalgebra $D(n, F)$ of diagonal matrices. Hence every element of the F-algebra $H \otimes F$ is diagonalizable. By definition every element of H is separable over K. In particular, H is semisimple because no non-zero fully separable element over K is nilpotent. H is solvable and thus H is torus. H is self-centralizing because it is commutative and self-normalizing: $H \subseteq C_A(H) = C_{A^\circ}(H) \subseteq N_{A^\circ}(H) = H$. For every commutative subalgebra C of A containing H the statement $C \subseteq C_A(C) \subseteq C_A(H) = H$ is valid.⋄

Theorem 33 *Let A be a central-simple finite-dimensional associative K-algebra. The following statements are valid:*

(i) The maximal tori of A are exactly the self-centralizing tori of A. In particular, every maximal torus of A is a maximal commutative and separable subalgebra of A.

(ii) The Cartan subalgebras of $A°$ are exactly the maximal tori of A.

(iii) Every Cartan subalgebra of $A°$ is $ind(A)$-dimensional.

(iv) The Cartan subalgebras of $A°$ are isomorphic.

Proof. ad(i): Let T be a maximal torus of A. Theorem 1 in [59] implies that $C_A(T)$ is a Cartan subalgebra of $A°$. By using lemma 7 this subalgebra is a torus, and the commutativity of T implies $T = C_A(T)$. Let C be a commutative subalgebra of A containing T. Then the following identity is valid: $C \subseteq C_A(C) \subseteq C_A(T) = T$. T is separable by lemma 4.

ad(ii): This statement is a consequence of part (i) and theorem 1 in [59].

ad(iii): see lemma 7.

ad(iv): All Cartan subalgebras of $A°$ are abelian (see part (ii)) and have by part (iii) the same K-dimension. ◇

Examples 2 (i) Let A be a central-simple finite-dimensional associative K-algebra. Every strict maximal subfield (which are subfields of dimension equal to the index of A) of A, which is separable over K, is a Cartan- subalgebra of $A°$: obviously such subfields are tori, and they are contained in a maximal torus. By using theorem 33) both tori are identical due to a dimension argument.

(ii) Let K be a field and $n \in \mathbb{N}$. The set of $D(n, K)$ of diagonal matrices over K is a Cartan subalgebra of $gl(n, K)$. Within the proof of lemma 7 we have used that for algebraically closed K its conjugates under $GL(n, K)$ are exactly the Cartan subalgebras of $gl(n, K)$.

(iii) Let $A := \mathbb{R}^{2 \times 2}$. By using theorem 33 we have to determine those self-centralizing subalgebras which are isomorphic to $\mathbb{C}$ or to $\mathbb{R} \times \mathbb{R}$.
Subalgebras of the first kind are strict maximal, and by part (i) they are Cartan subalgebras of $A°$. Let $M \in A$, like $\begin{pmatrix} a_1 & a_2 \\ a_3 & a_4 \end{pmatrix}$, with $M^2 = -1$.
The following conditions are valid for M:

$$\begin{aligned} a_1^2 + a_2 a_3 &= -1 \quad (1) \\ a_2(a_1 + a_4) &= 0 \quad (2) \\ a_3(a_1 + a_4) &= 0 \quad (3) \\ a_4^2 + a_2 a_3 &= -1 \quad (4). \end{aligned}$$

Squares are non-negative, hence a_2 and a_3 are by using condition (4) non-zero. For arbitrary a_1 and $a_2 \neq 0$ the conditions (2) and (3) imply $a_3 = \frac{-(1+a_1^2)}{a_2}$ and $a_4 = -a_1$. Vice versa, it is straightforward to calculate that the square of the matrix $\begin{pmatrix} a_1 & a_2 \\ \frac{-(1+a_1^2)}{a_2} & -a_2 \end{pmatrix}$ is exactly -1.

Subalgebras isomorphic to $\mathbb{R} \times \mathbb{R}$ are of the form $\mathbb{R}[t]$, in which t is an idempotent. Idempotents are diagonalizable. Hence there exists an element $g \in E(A)$ with $t^g \in D(2, \mathbb{R})$. As a consequence such subalgebras are conjugate to the subalgebra of diagonal matrices which are by part (ii) Cartan subalgebras of A°. ⋄

Example 4 *(cyclic algebras)* Within this example we focus on so-called cyclic algebras which are central-simple associative algebras and generalizing quaternion algebras. An introduction is contained (including proofs) in [39]. Let $(K; L)$ be a Galois extension of degree n with cyclic Galois group generated by σ. On the additive group of the polynomial algebra $L[t]$ we define a new composition by $at^i \Delta bt^j := a(b\sigma^i)t^{i+j}$ for all $a, b \in L$ and $i, j \in \mathbb{N}_0$. As K is contained in the fixed field of σ we get by the distributive extension of this multiplication on $L[t]$ an associative algebra denoted by $L[t, \sigma]$.

For an element $z \in K$ with $b \neq 0$ let $L_{\sigma,z}$ be the factor algebra by the ideal generated by $t^n - z$ in $L[t, \sigma]$, and every algebra isomorphic to such an factor algebra is called a cyclic algebra.

Cyclic algebras are of dimension n^2 and they are centra-simple and associative. It can be proven that the naturally embedded extension field L is self-centralizing in the cyclic algebra. Its dimension is n – which is the index of the cyclic algebra — and thus the field is strict maximal and separable. By using the examples 2, part (i), we conclude that it is a Cartan subalgebra and a maximal torus.

Cyclic algebras are generalizing quaternion algebras and one can derive division algebras of higher dimensions. In [39] a sufficient condition is given for this statement: if n a prime number and the cyclic algebra non-isomorphic to $K^{n \times n}$, then its a central division algebra. The condition not being isomorphic to $K^{n \times n}$ is analyzed by the following criteria: the cyclic algebra is isomorphic to $K^{n \times n}$ if and only if an element $b \in L$ exists such that $z = \prod_{i=0}^{n-1}(b\sigma)^i$ is valid. In the exercises a concrete example is to be calculated by the reader which is also included in [39]. ⋄

5.7.2 Simple algebras

A simple finite-dimensional associative algebra A is central as $Z(A)$-algebra. Every Cartan subalgebra of A° is containing the center of A and is by using proposition 8 a $Z(A)$-subalgebra of A. Therefor theorem 33 is used to prove the following theorem:

Theorem 34 *Let A be a simple and finite-dimensional associative K-algebra. The Cartan subalgebras of A° are exactly the unital commutative subalgebras of A, which are maximal with the condition that every element is fully separable over the center of A. These subalgebras are self-centralizing and maximal commutative as associative subalgebras.*

In particular, every Cartan subalgebra T of A° is a direct sum of fields and of dimension $\dim_K(Z(A)) \cdot \operatorname{ind}_{Z(A)}(A)$. All Cartan subalgebras of A° are isomorphic.⋄

By using theorems 33 and 34 we deduce the following corollary for separable simple algebras:

Corollary 3 *Let A be a simple finite-dimensional associative K-algebra, such that its center is separable over K.*

(i) *The maximal tori of A are exactly the self-centralizing tori of A. In particular, every maximal torus of A is a maximal commutative and separable subalgebra of A.*

(ii) *The Cartan subalgebras of A° are identical with the maximal tori of A.*

(iii) *Every Cartan subalgebra T of A° is of dimension $\dim_K(Z(A)) \cdot \operatorname{ind}_{Z(A)}(A)$.*

(iv) *All Cartan subalgebras of A° are isomorphic.*⋄

5.7.3 Semisimple and separable algebras

In this section we reduce the analysis of Cartan subalgebras of associated Lie algebras of semisimple associative algebras to their simple components.

Let A, B be associative K-algebras. For all $a_1, a_2 \in A$ and $b_1, b_2 \in B$ the identity $(a_1; b_1) \circ (a_2; b_2) = (a_1 \circ a_2; b_1 \circ b_2)$ is valid.[5] By this rule we derive for $T \subseteq A$ and $S \subseteq B$ the equation $N_{(A \times B)^\circ}(T \times S) = N_{A^\circ}(T) \times N_{B^\circ}(S)$, and we are ready to prove the following remark:

---

[5] $\circ$ is defined componentwise on the direct product.

Remark 11 (i) Let $n \in \mathbb{N}$, $A_1, \cdots, A_n$ be associative K-algebras and $C_1, \cdots, C_n$ Cartan subalgebras of $(A_1)^\circ, \cdots, (A_n)^\circ$, then $C_1 \times \cdots \times C_n$ is a Cartan subalgebra of $(A_1 \times \cdots \times A_n)^\circ$.

(ii) Let A, B be associative finite-dimensional K-algebras and C a Cartan subalgebra of $(A \times B)^\circ$. We define $T := \{a \mid a \in A, \exists b \in B : (a,b) \in C\}$ and $S := \{b \mid b \in B, \exists a \in A : (a,b) \in C\}$. T resp. S is a nilpotent subalgebra of A° resp. B°. In particular, $T \times S$ is a nilpotent subalgebra of $(A \times B)^\circ$ containing by definition C as a subalgebra. As C is maximal nilpotent, we conclude $C = T \times S$. By using the identity $T \times S = C = N_{(A \times B)^\circ}(C) = N_{(A \times B)^\circ}(S \times T) = N_{A^\circ}(T) \times N_{B^\circ}(T)$ we derive that T resp. S is a Cartan subalgebra of A° resp. B°.⋄

We apply the previous remark and derive the following reduction theorem:

Theorem 35 *Let $n \in \mathbb{N}$, $A_1, \cdots, A_n$ finite-dimensional associative K-algebras. The Cartan subalgebras of $(A_1 \times \cdots \times A_n)^\circ$ are exactly those associative subalgebras $C_1 \times \cdots \times C_n$ that for every $i \in \underline{n}$ the set C_i is a Cartan subalgebra of $(A_i)^\circ$.*⋄

For separable algebras we deduce by corollary 3, theorem 35 and theorem 1 in [59]:

Theorem 36 *Let A be a finite-dimensional associative separable K-Algebra.*

(i) Every maximal tori of A is a direct sum of maximal tori of the simple ideals decomposing A direct.

(ii) The maximal tori of A are exactly the self-centralizing tori of A. In particular, the maximal tori are maximal commutative, separable subalgebras of A.

(iii) The Cartan subalgebras of A° are exactly the maximal tori of A.

(iv) All Cartan subalgebras of A° are isomorphic.

(v) The dimension of the Cartan subalgebras of A° is the sum of the dimensions of the Cartan subalgebras of the Lie algebras associated to the simples ideals of A decomposing A directly.⋄

Let $r, n_1, \ldots, n_r \in \mathbb{N}$ and $D_1, \ldots, D_r$ division algebras such that A is isomorphic to $\bigoplus_{i=1}^{r} D_i^{n_i \times n_i}$. The maximal tori of A and the Cartan subalgebras of A° are of the unique dimension

$$\sum_{i=1}^{r} dim_K(Z(D_i)) \cdot ind_{Z(D_i)}(D_i^{n_i \times n_i}) = \sum_{i=1}^{r} n_i \cdot dim_K(Z(D_i)) \cdot ind_{Z(D_i)}(D_i).$$

Section 5.7

Central-Simple
$\bar{J}_A = \mathbb{C}_{A^0}$
maximal tori
=
self-centralizing tori

Cartan subalgebras
{
• ad(D)-diagonal
• separable
• maximal commutative
• all are isomorphic
}

Simple
$\bar{J}_A = \mathbb{C}_{A^0}$
maximal tori
=
self-centralizing tori

Cartan subalgebras
{
• ad(a), ad(A)-diagonal
• separable
• $\mathfrak{h}$, isomorphic
}

Semi-Simple
$\bar{J}_A = \mathbb{C}_{A^0}$
direct products of components
no diagonals
maximal tori
Cartan subalgebras

5.7.4 Open-ended questions and exercises

Open-ended questions 6 *(i) Let $n \in \mathbb{N}$ and K be a field. For an arbitrary field K or for special cases of K (finite or $K \in \{\mathbb{Q}, \mathbb{R}\}$) determine the maximal tori and Cartan subalgebras of $gl(n, K)$.*

(ii) What are the maximal tori (= Cartan subalgebras) of cyclic algebras?

(iii) What are the maximal tori (= Cartan subalgebras) of $D^{n \times n}$ for a (separable) division algebra D and $n \in \mathbb{N}$?

Excercise 162 *Study the content of [91] dealing with the determination of Cartan subalgebras of $D^{n \times n}$ for a division algebra D and $n \in \mathbb{N}$.*

Excercise 163 *Prove that maximal commutative subalgebras are self-centralizing.*

Excercise 164 *Determine the maximal commutative subalgebras of a finite-dimensional division algebra. Do they have all the same dimension? Solve this exercise also for a central-division algebra.*

Excercise 165 *Let A be a finite-dimensional associative simple algebra. Prove that the Cartan subalgebras of $A°$ are maximal commutative subalgebras of A.*

Excercise 166 *Study the article of N. Jacobson [30]! For a field K and a natural number n let $N(n)$ be the maximal dimension of all commutative subalgebras of $K^{n \times n}$. If $n = 2g$ is even, then the identity $N(n) = g^2 + 1$ is valid. In the case $n = 2u + 1$ the formula $N(n) = u(u-1) + 1$ is valid. Is the dimension of the Cartan subalgebras and upper or lower bound for $N(n)$? Does a number n exist such that $N(n)$ is exactly the dimension of a Cartan subalgebra? For $n \leq 20$ present the numbers $N(n)$ in table form.*

Excercise 167 *Let K be a field and $n \in \mathbb{N}$. Is the dimension of all maximal commutative subalgebras of $K^{n \times n}$ identically? (Tip: exercise 166)*

Excercise 168 *Let K be a field and G a finite group. If KG is modular, then prove that $rad(KG)$ is of dimension 1 if and only if $rad(KG) = \langle \overline{G} \rangle_K$ is valid. In particular, $rad(KG)$ is central in this case.*

Excercise 169 *True or false: The maximal dimension of commutative subalgebras is the maximal dimension of all maximal commutative subalgebras of a finite dimensional associative algebra.*

Excercise 170 *Let $n \in \mathbb{N}$ and D a finite-dimensional associative central K-division algebra. Use the theorem that for a maximal subfield T of D the tensor product $D \otimes T$ is isomorphic to $T^{r \times r}$ as T-algebras. The value r is the* well-known index of D – denoted by $ind_K(D)$ – which is the square root of $dim_K(D)$. If A is a (maximal) commutative subalgebra of $D^{n\times n}$, then – by using the results of exercise 166 – the inequality $dim_K(A) \leq N(n \cdot ind_K(D))$ is valid. (additional tip: exercise 166 and T is a splitting field for $D^{n\times n}$)

Excercise 171 We focus again on the exercises 166 and 170. Use the article of N. Jacobson in [30] to analyze whether the presented bound is met or not. What is the meaning of Cartan subalgebras for this exercise?

Excercise 172 Within the exercises 170 and 171 let D be not necessary central. D is a central algebra over the field $Z(D)$ and every maximal commutative subalgebra is containing the center. An upper bound for their maximal dimension is $dim_K(Z(D)) \cdot N(n \cdot ind_{Z(D)}(D))$. Use again the article of Jacobson ([30]) to decide whether the bound is met.

Excercise 173 Do research in literature to find an article of Malcev extending the results of Jacobson [30] to commutative subalgebras of simple complex Lie algebras (Tip: commutative subalgebras of semi-simple Lie algebras). Present the results of Malcev in table form. (Tip: Within the article which is written in Russian there is an English summary!).

Excercise 174 Is it possible to transfer all results of theorem 36 by replacing separable by semisimple?

Excercise 175 Is every quaternion algebra cyclic? Is the opposite statement true?

Excercise 176 The real quaternion algebra is cyclic with respect to the complex conjugation and $z = -1$ (Tip: see [39])!

Excercise 177 By studying [39] prove all statements of example 4!

Excercise 178 By using the results presented in example 4 prove the following: If L is a splitting field for the polynomial $t^3 + t^2 - 2t - 1 \in \mathbb{Q}[t]$ in $\mathbb{C}$ and $b \in L$ a root of this polynomial, then the Galois group of $(\mathbb{Q}, L)$ is cyclic and generated by the automorphism σ mit $b\sigma = b^2 - 2$ of order 3. In addition, prove that no $b \in L$ exists such that $\prod_{i=0}^{3-1}(b\sigma)^i = 2$ is valid. Deduce that $L_{\sigma,2}$ is a central division algebra of dimension 9 over $\mathbb{Q}$ (Tip: see [39]).

Excercise 179 Determine all Cartan subalgebras and maximal tori of $gl(3,\mathbb{R})$!

Excercise 180 Let $n \in \mathbb{N}$. Determine all Cartan subalgebras and maximal tori of $gl(n,\mathbb{C})$!

Excercise 181 Determine all Cartan subalgebras and maximal tori of $gl(2,\mathbb{Q})$!

Excercise 182 *Let K be a field with p elements. Determine all Cartan subalgebras and maximal tori of $gl(2, K)$?*

Excercise 183 *Let $n \in \mathbb{N}$. Determine all Cartan subalgebras and maximal tori of A° for the following cases of A:*

(i) $A = \mathbb{R} \times \mathbb{C}$ as $\mathbb{R}$-Algebra

(ii) $A = \mathbb{R}^{2\times 2} \times \mathbb{R}$ as $\mathbb{R}$-algebra

(iii) $A = \mathbb{R}^{2\times 2} \times \mathbb{C}$ as $\mathbb{R}$-algebra

(iv) $A = \mathbb{C}^{n\times n} \times \mathbb{C}$ as $\mathbb{C}$-algebra

Excercise 184 *Construct a non-cyclic division algebra! (Tip: [84])*

Excercise 185 *On what terms are the maximal tori and Cartan subalgebras are identical for the Lie algebra associated to a simple or semisimple finite-dimensional associative algebra?*

Excercise 186 *Reduce the determination of maximal nilpotent Lie subalgebras of direct products to the corresponding ones of their factors.*

Excercise 187 *In chapter 1 of [74] two algebras are presented without having a radical complement. Determine the maximal tori and Cartan subalgebras of the semisimple factor algebra by its nilradical. Is every maximal tori a Cartan subalgebra and vice versa?*

Excercise 188 *(zero-extension) Analyze within exercise 39 for the zero-extension on what terms the factor algebra by its nilradical is central-simple, simple, semisimple or separable. Determine the maximal tori and Cartan subalgebras and connect the results to the original algebra the zero-extension is based on. If the analysis is to complex, then assume e.g. that the field is perfect.*

Excercise 189 *(eAe) Let A be a finite dimensional associative unitary algebra and e an idempotent of A. Analyze within exercise 136 for eAe on what terms the factor algebra by its nilradical is central-simple, simple, semisimple or separable. Determine the maximal tori and Cartan subalgebras and connect the results to the original algebra A. If the analysis is to complex, then assume e.g. that the field is perfect or the idempotent e is central.*

5.8 Basic algebras

In the case of basic – or also called reduced – finite-dimensional associative algebras we can transfer the determination of Cartan subalgebras to the corresponding determination of special maximal solvable subalgebras. We finish this section with a characterization of reduced algebras by maximal nilpotent subalgebras. In addition, we determine condition for the group algebra being reduced and calculate the dimension of their Cartan subalgebras for basic algebras.

5.8.1 Characterizations of reduced algebras

Definition and remark 4 If A is an associative K-algebra, then we denote by $nil(A)$ the set of all nilpotent elements of A. The nilradical of A is contained in $nil(A)$. We call A reduced or basic, if $rad(A) = nil(A)$ is valid.◇

Proposition 15 Let A be a finite-dimensional associative K-algebra. The following conditions are equivalent:

(i) A is reduced.

(ii) $A/rad(A)$ is reduced.

(iii) $A/rad(A)$ is a direct sum of K-division algebras.
    In particular, A reduced if A is commutative or solvable.

(iv) For every subalgebra T of A the condition $rad(T) = rad(A) \cap T$ is valid.

(v) Every subalgebra of A is reduced.

(vi) For every ideal I of A the factor algebra A/I is reduced.

Proof. We remark the following statement:
(*) If $n \in \mathbb{N}$ and D is a division algebra, then $D^{n \times n}$ possesses an nilpotent element different from zero if and only if $n \geq 2$ is valid.

(i) $\to$ (ii): Let $a \in A$ such that $a + rad(A)$ is nilpotent. Then there exists an element $n \in \mathbb{N}$ with $a^n \in rad(A)$. We conclude that a is nilpotent. By using (i) we obtain $a \in rad(A)$, and we have proven $a + rad(A) = 0 + rad(A)$.

(ii) $\to$ (iii): This is a direct consequence of (*).

(iii) $\to$ (i): Let a be a nilpotent element of A. Obviously, $a + rad(A)$ is nilpotent. By using (*) we conclude $a + rad(A) = 0 + rad(A)$.

(i) → (iv): Let T be a subalgebra of A. By using condition (i) we derive $rad(T) \subseteq nil(T) \subseteq nil(A) = rad(A)$, and hence $rad(T) \subseteq rad(A) \cap T$ is valid. $rad(A) \cap T$ is a nilpotent ideal of T, and we have proven (iv).

(iv) → (v): Let T be a subalgebra of A and $t \in nil(T)$. We define the commutative subalgebra $S := K[t]$ of T. By using the equivalence of (iii) and (i) we conclude that S is reduced, and $rad(S) = nil(S)$ is valid. By using condition (iv) we obtain $t \in nil(S) = rad(S) = rad(A) \cap S \subseteq rad(A) \cap T \subseteq rad(T)$.

(v) → (i): This statement is straightforward to prove.

(vi) → (i): This statement is straightforward to prove, too.

(i) → (vi): Let I be an ideal of A and A reduced. The radical of A/I is well-known: $(rad(A) + I)/I$. The factor structure by this radical is isomorphic to $A/(rad(A) + I)$, and this one is isomorphic to a factor algebra of the semisimple reduced algebra $A/rad(A)$. This ideal of the semisimple algebra possesses an ideal complement. Thus the structure to be analyzed is isomorphic to an ideal of a semisimple reduced algebra. By using the equivalence of (v) and (i) the proof is finished.⋄

5.8.2 Maximal solvable subalgebras

By using lemma 4 the following remark is valid:

Remark 12 Let A a finite-dimensional associative unitary K-algebra, T a torus of A and $S := T \oplus rad(A)$. Then T is a separable radical complement of $rad(S) = rad(A)$ in S, and S is a solvable subalgebra of A.⋄

Lemma 8 Let A be an associative finite-dimensional unitary reduced K-algebra. Upon the subalgebras of A with separable factor algebra by their nilradical the maximal solvable elements can be presented by $rad(A) \oplus T$, T a maximal torus of A.

Proof. Let T be a maximal torus of A and $S := rad(A) \oplus T$. Because of remark 12 the subalgebra S of A is solvable and possesses a separable factor algebra by its nilradical. Let B be a solvable subalgebra of A possessing a separable factor algebra by its nilradical and containing $rad(A) \oplus T$. A is reduced, and we derive by using proposition 15 the identity $rad(B) \subseteq rad(A)$. As a torus T is – with respect to lemma 4 – a separable subalgebra of B, which is contained by corollary 2.3.7 in [74] in a radical complement X of B. X is – by using lemma 4 – a torus of A, and we conclude $T = X$ by the maximality of T. Hence $B = rad(A) \oplus T$ is valid.

Let B be upon subalgebras of A with separable factor algebra by their nilradical a maximal solvable element and T a radical complement of B.

Then T is with respect to lemma 4 a torus of A, and proposition 15 implies $rad(B) \subseteq rad(A)$. We define $S := rad(A) \oplus T$. By remark 12 and the maximality of B we deduce $B = S$ and $rad(B) = rad(A)$. If R would be a torus that contains T proper, then the subalgebra $rad(A) \oplus R$ would be containing B proper, too. By using remark 12 we would derive a contradiction to the maximality of B.⋄

By using lemma 8 we derive:

Corollary 4 *Let A be an associative finite-dimensional unitary reduced K-algebra. If K is perfect, then the maximal solvable subalgebras of A can be presented by $T \oplus rad(A)$, T being a maximal torus of A.*⋄

Counterexample 2 Let K be a field and $n \in \mathbb{N}$, then the algebra of lower triangular matrices is a solvable and non-commutative subalgebra of the semisimple associative algebra $K^{n \times n}$. Thus corollary 4 fails for $K^{n \times n}$.⋄

5.8.3 Cartan subalgebras

Remark 13 Let A be an associative K-algebra, I an ideal and T a subalgebra of A, such that A is the inner direct sum of I and T. For every subset X of T the identity $C_A(X) = C_I(X) \oplus C_T(X)$ is valid.⋄

Theorem 37 *Let A be a finite-dimensional associative unitary reduced K-algebra with separable factor algebra by its nilradical and $\mathcal{S}(A)$ the set of solvable subalgebras of A possessing a separable factor algebra by its nilradical. The Cartan subalgebras of $A°$ correspond with the Cartan subalgebras of Lie algebras associated to maximal elements of $\mathcal{S}(A)$.*

Proof. Let H be a Cartan subalgebra of $A°$. By using theorem 1 in [59] a maximal torus T of A exists such that $H = C_A(T)$ is valid. We define $S := rad(A) \oplus T$. Lemma 8 implies that S is a maximal element of $\mathcal{S}(A)$. By using 2.3.7 in [74] and lemma 4 a radical complement C of A exists containing T. T is a maximal torus of C, and by using theorem 36 its self-centralizing in C. Remark 13 implies $H = C_A(T) = C_{rad(A)}(T) \oplus T$. T is a maximal torus of S, too. By using the main theorem 1 the subalgebra $C_S(T)$ is a Cartan subalgebra of $S°$. T is commutative, and hence – by using remark 13 – the identity $C_S(T) = C_{rad(A)}(T) \oplus T = C_A(T) = H$ is valid. We derive that H is a Cartan subalgebra of S.

Let S be a maximal element of $\mathcal{S}(A)$. Lemma 8 implies the existence of a maximal torus T of A such that $S = rad(A) \oplus T$ is valid. Every Cartan subalgebra of S is – with respect to theorem 24 – the centralizer of a radical complement of S. The theorem of Wedderburn-Malcev provides that these complements are exactly the conjugates of T under $1 + rad(S) = 1 + rad(A)$. Let H a Cartan subalgebra of S, and we assume w.l.o.g that $H = C_S(T)$ is valid.

Remark 13 and the commutativity of T imply $H = C_S(T) = C_{rad(A)}(T) \oplus T$. By using the main theorem 1 the subalgebra $C_A(T)$ is a Cartan subalgebra of $A°$. Let C be a radical complement of A containing T (see 2.3.7 in [74] and lemma 4). With respect to remark 13 and theorem 36 we conclude $C_A(T) = C_{rad(A)}(T) \oplus C_C(T) = C_{rad(A)}(T) \oplus T = C_S(T) = H$.⋄

A direct consequence of theorem 37 is the following corollary:

Corollary 5 *Let A be an associative finite-dimensional unitary reduced K-algebra. If K is perfect, then the Cartan subalgebras of $A°$ are exactly Cartan subalgebras of Lie algebras associated to maximal solvable subalgebras of A.*⋄

Remark 14 Let A be a finite-dimensional associative unitary reduced K-algebra with separable factor algebra by its nilradical. In view of theorem 37 it would be preferable if every solvable subalgebra possesses a separable factor algebra by its nilradical. For a separable K-division algebra this statement is equivalent to the result that every subfield is separable. In section 5.4.1 **Quaternion algebras in characteristics** 2 we have proven that a non-separable maximal subfield exists.⋄

5.8.4 Maximal nilpotent subalgebras

Proposition 16 *Let D be a division algebra and $n \in \mathbb{N}_{\geq 2}$. Then an idempotent $\neq 0$ in $D^{n \times n}$ exists which is the sum of two nilpotent elements $\neq 0$.*

Proof. We focus on the matrices $A := (a_{i,j})$, $B := (b_{i,j})$ and $C := (c_{i,j})$ such that A, B possesses only the non-zero entries $a_{1,2} = b_{2,1} = 1$. A, B are nilpotent and non-zero, and $C := A + B$ is an idempotent different from zero in $D^{n \times n}$.⋄

Proposition 17 *Let A be an associative finite-dimensional algebra with exactly one maximal nilpotent subalgebra T. Then $(T + rad(A))/rad(A)$ is the only maximal nilpotent subalgebra of $A/rad(A)$.*

Proof. T is nilpotent and thus $(T + rad(A))/rad(A)$ is nilpotent, too. Let X be a nilpotent subalgebra of $A/rad(A)$. Let M be a subalgebra of A such that $X = M/rad(A)$ is valid. $M/rad(A)$ is nilpotent and thus a power of M is contained in $rad(A)$. By using the nilpotency of $rad(A)$ we derive that M is nilpotent, too. T is the only maximal nilpotent subalgebra and we conclude $M \subseteq T$ and by this $X \subseteq (T + rad(A))/rad(A)$. ⋄

Theorem 38 *Let A be a finite-dimensional associative algebra. A is reduced if and only if A possesses exactly one maximal nilpotent subalgebra. In this case $rad(A) = nil(A)$ is the unique maximal nilpotent subalgebra.*

Proof. Let A be reduced, and by definition $rad(A) = nil(A)$ is valid. The nilradical contains every nilpotent element and thus every nilpotent subalgebra, too. The radical is also a nilpotent subalgebra and therefor its the only maximal nilpotent subalgebra.

Let T be the only maximal nilpotent subalgebra of A. Then $A/rad(A)$ possesses – by using proposition 17 – exactly one maximal nilpotent subalgebra which is exactly $(T + rad(A))/rad(A)$. This nilpotent subalgebra contains every nilpotent element because every nilpotent element generates a nilpotent subalgebra. If $A/rad(A)$ is not reduced, then $A/rad(A)$ would contain – by using proposition 16 – a non-zero idempotent which is the sum of two nilpotent non-zero elements. This sum would be contained in the maximal subalgebra, and this subalgebra would contain a non-zero idempotent. This idempotent would be nilpotent as well and hence it is zero. This is a contradiction. Hence we conclude that A is reduced.⋄

Within the next section we answer the following questions: On what terms is the group algebra KG reduced and what information can we derive for the maximal tori and Cartan subalgebras of $(KG)^\circ$?

5.8.5 Reduced group algebras: the semisimple case

By our previous results we know that maximal tori (= Cartan-subalgebras, because a semisimple group algebra is separable) resp. the nilradical are direct products of maximal tori resp. nilradicals of the ideals which are decomposing KG directly. In addition, if the group algebra is reduced, then these ideals are division algebras. We analyze which division algebras play a decisive role.

Proposition 18 *Let G be a finite group, N a subgroup of G, K a field such that KG is semisimple, $\overline{N} := \sum_{n \in N} n$ and $e_N := \frac{1}{|N|_K} \cdot \overline{N}$. e_N is an idempotent of KG which central if and only if N is a normal subgroup of G.*

Proof. N is a subgroup and thus $NN = N$ is valid. We conclude that e_N is an idempotent. In addition, it is straightforward to prove that $N^g = N$ for all $g \in G$ is valid if and only if e_N is central.⋄

Dedekind-groups are defined by the condition that every subgroup is normal. The non-abelian Dedekind-groups are called hamiltonian groups.

Corollary 18 *Let G be a non-abelian finite group and K a field such that KG is semisimple and reduced. Then G is hamiltonian.*[6]

---

[6]Sir William Rowan Hamilton (born midnight, 3rd to 4th August 1805, died 2nd September 1865) was an Irish physicist, astronomer, and mathematician, who made important contributions to classical mechanics, optics, and algebra. His studies of mechanical

Proof. By our assumptions KG is a direct sum of division algebras, like $D_1 \times \cdots \times D_n$. Let N be a subgroup of G. We focus on the idempotent e_N defined in proposition 18. For every $i \in \underline{n}$ there are elements $d_i \in D_i$ with $e_N = (d_1, \cdots, d_n)$. e_N is idempotent, and hence every d_i, $i \in \underline{n}$ is idempotent, too. We conclude $d_i \in \{1, 0\}$ for every i, because each D_i is a division algebra. As a consequence we derive $d_i \in Z(D_i)$ for all i, and thus e_N is central in KG. By using proposition 18 the proof is complete. ⋄

Hamiltonian groups are classified by Richard Dedekind[7] (see [13]): $Q_8 \times$

---

and optical systems led him to discover new mathematical concepts and techniques. His best known contribution to mathematical physics is the reformulation of Newtonian mechanics, now called Hamiltonian mechanics. This work has proven central to the modern study of classical field theories such as electromagnetism, and to the development of quantum mechanics. In pure mathematics, he is best known as the inventor of quaternions. Hamilton is said to have shown immense talent at a very early age. Astronomer Bishop Dr. John Brinkley remarked of the 18 year old Hamilton: This young man, I do not say will be, but is, the first mathematician of his age. William Rowan Hamilton's scientific career included the study of geometrical optics, classical mechanics, adaptation of dynamic methods in optical systems, applying quaternion and vector methods to problems in mechanics and in geometry, development of theories of conjugate algebraic couple functions (in which complex numbers are constructed as ordered pairs of real numbers), solvability of polynomial equations and general quintic polynomial solvable by radicals, the analysis on Fluctuating Functions (and the ideas from Fourier analysis), linear operators on quaternions and proving a result for linear operators on the space of quaternions (which is a special case of the general theorem which today is known as the Cayley-Hamilton theorem). Hamilton also invented icosian calculus, which he used to investigate closed edge paths on a dodecahedron that visit each vertex exactly once.

[7]Julius Wilhelm Richard Dedekind (born 6th October 1831, died 12nd February 1916) was a German mathematician who made important contributions to abstract algebra (particularly ring theory), algebraic number theory and the definition of the real numbers. Dedekind's father was Julius Levin Ulrich Dedekind, an administrator of Collegium Carolinum in Braunschweig. Dedekind had three older siblings. As an adult, he never used the names Julius Wilhelm. He was born, lived most of his life, and died in Braunschweig (often called "Brunswick" in English). He first attended the Collegium Carolinum in 1848 before transferring to the University of Göttingen in 1850. There, Dedekind was taught number theory by professor Moritz Stern. Gauss was still teaching, although mostly at an elementary level, and Dedekind became his last student. Dedekind received his doctorate in 1852, for a thesis titled Über die Theorie der Eulerschen Integrale ("On the Theory of Eulerian integrals"). This thesis did not display the talent evident by Dedekind's subsequent publications. At that time, the University of Berlin, not Göttingen, was the main facility for mathematical research in Germany. Thus Dedekind went to Berlin for two years of study, where he and Bernhard Riemann were contemporaries; they were both awarded the habilitation in 1854. Dedekind returned to Göttingen to teach as a Privatdozent, giving courses on probability and geometry. He studied for a while with Peter Gustav Lejeune Dirichlet, and they became good friends. Because of lingering weaknesses in his mathematical knowledge, he studied elliptic and abelian functions. Yet he was also the first at Göttingen to lecture concerning Galois theory. About this time, he became one of the first people to understand the importance of the notion of groups for algebra and arithmetic. In 1858, he began teaching at the Polytechnic school in Zürich (now ETH Zürich). When the Collegium Carolinum was upgraded to a Technische Hochschule (Institute of Technology) in 1862, Dedekind returned to his native Braunschweig, where he

$A \times (Z_2)^n$, A an abelian group of odd order (also $|A| = 1$ is possible) and $n \in \mathbb{N}_0$. Abelian groups result in nilpotent associated Lie group algebras. Therefor we are focussing on hamiltonian groups only and clarify for which hamiltonian groups a semisimple group algebra is reduced. The next lemma summarizes some well-known facts which will be used to study this question.

Lemma 9 *Let K be a field, F a field extension of K, Q a quaternion algebra over K, H a non-abelian hamiltonian group, $n \in \mathbb{N}_0$, G, M finite groups, A a finite abelian group of order r, for every divisor d of r the element ω_d a primitive d-th root of unity and $a_d := \frac{|\{a|a \in A, o(a) = d\}|}{dim_K(K(\omega_d))}$. The following statements are valid:*

(i) *If $char(K) \neq 2$ is valid, then $K(Z_2)^n$ and K^{2^n} are isomorphic.*

(ii) *If KA is semisimple, then KA is isomorphic to $\bigoplus_{d|r} K(\omega_d)^{a_d}$.*

(iii) *Let KQ_8 be semisimple. If K is a splitting field for Q_8, then $KQ_8 \cong K^4 \oplus K^{2 \times 2}$ is valid.*

(iv) *Let KQ_8 be semisimple. If K is no splitting field for KQ_8, then there exists a quaternion algebra B over K which is a division algebra such that so $KQ_8 \cong K^4 \oplus B$ is valid.*

(v) *Let KQ_8 be semisimple. K is no splitting field for KQ_8 if and only if the equation $a^2 + b^2 + 1 = 0$ has no solution over K.*

(vi) *If KH is semisimple, then $char(K) \neq 2$ is valid.*

(vii) *$B \otimes F$ is a quaternion algebra over F.*

(viii) *$K(G \times M) \cong KG \otimes KM$.*

Proof. ad(i): Every element x of the elementary abelian 2-group is diagonalizable: $x^2 = 1$ is valid, and thus the minimal polynomial of x is $(t+1)(t-1)$. By basic results in [74] this part is proven.

ad(ii): see [48]

ad(iii)-(v): see e.g. [39]

ad(vi): This is a consequence of a theorem of Maschke.

---

spent the rest of his life, teaching at the Institute. He retired in 1894, but did occasional teaching and continued to publish. He never married, instead living with his sister Julia. Dedekind was elected to the Academies of Berlin (1880) and Rome, and to the French Academy of Sciences (1900). He received honorary doctorates from the universities of Oslo, Zurich, and Braunschweig.

ad(vii): see e.g. [49]

ad(viii): a straightforward calculation.◇

Theorem 39 *Let K be a field and G a finite hamiltonian group such that KG is semisimple. Then a finite abelian group A of odd order r and $n \in \mathbb{N}_0$ exist such that $G \cong Q_8 \times A \times (Z_2)^n$ is valid. KG is reduced if and only if the equation $a^2 + b^2 + 1 = 0$ has no solution in every field extension $(K; K(\omega_d))$ for $d \mid r$ and primitive dth root of unity. In addition, $char(K) = 0$ is valid.*

Proof. We use some well-known properties of the tensor product: it is distributive with respect to the direct sum and does not change by tensoring with the base field. In addition, we use the results of lemma 9. Our aim is to decompose the semisimple group algebra in their simple components. Let $\mid A \mid = r$, for every divisor d of r the element ω_d a primitive dth root of unity and $a_d := \frac{|\{a|a \in A, o(a)=d\}|}{dim_K(K(\omega_d))}$. KG is isomorphic to the tensor product $K(Q_8) \otimes KA \otimes (K(Z_2))^n$. Hence there is a quaternion algebra Q_K over K, such that KG is isomorphic to $(K^4 \oplus Q_K) \otimes (\bigoplus_{d|r} (K(\omega_d))^{a_d}) \otimes (KZ_2)^n$. For all $d \mid r$ let Q_{K_d} be a quaternion algebra over the field extension $K(\omega_d)$ of K. We derive that KG is isomorphic to $(K^4 \oplus KA) \otimes (\bigoplus_{d|r} (Q_{K_d})^{a_d}) \otimes (KZ_2)^n$ and hence to $\bigoplus_{d|r}(K(\omega_d))^{4 \cdot a_d \cdot 2^n} \oplus \bigoplus_{d|r}(Q_{K_d})^{2^n \cdot a_d}$. We conclude that KG is decomposable into a direct product of division algebras if and only if the equation $a^2 + b^2 + 1 = 0$ has no solution in every field extension $(K; K(\omega_d))$ for $d \mid r$ and primitive dth root of unity

(sum of two squares theorem) We assume that $char(K) = p > 0$ is valid. The prime field of K is isomorphic to $P := GF(p)$ which is a finite field. Let P^2 the set of all squares of elements of P different to zero. Then $\mid P^2 \mid >= \frac{|P|-1}{2}$. Let F the set of all elements of P which are the sum of two squares of P, and let $F^* := F \setminus \{0\}$. We prove $F = K$. By $(a^2 + b^2)(c^2 + d^2) = (ac - bd)^2 + (ad + bc)^2$ the set F^* is a subgroup of the multiplicative group of P. F^* is containing P^2, which possesses an index $<= 2$ in P^*. Hence $F^* = P^*$ (and the proof is finished) is valid, or the identity $F^* = P^2$ is true. Then every sum of two squares is again a square. Hence F would be a subgroup of the additive group of P. As $\mid F \mid = \mid F^* \mid + 1 >= \frac{|P|+1}{2} > \frac{|P|}{2}$ is valid, we obtain by the theorem of Lagrange $F = P$.◇

Definition 2 *A K-algebra A is called reversible, if for all $a, b \in A$ with $ab = 0$ the condition $ba = 0$ is valid.*◇

Theorem 40 *Let K be a field and G a finite group, such that KG is semisimple. KG is reduced if and only if KG is reversible.*

Proof. Let KG be reduced and the sum of division algebras $D_1 \times \cdots \times D_n$. Let $(d_1, ..., d_n), (e_1, ..., e_n)$ two elements of KG such that the product $(d_1, ..., d_n)(e_1, ..., e_n)$ is equal to zero. Then we have $d_i e_i = 0$ for all $i \in \underline{n}$. Because there are no divisors of zero we derive $d_i = 0$ or $e_i = 0$. Hence we conclude $e_i d_i = 0$ for all $i \in \underline{n}$. As a consequence the product $(e_1, ..., e_n)(d_1, ..., d_n)$ is zero, and KG is reversible.

We have to prove that reversible group algebras are reduced. KG is a direct sum of matrix algebras over division algebras $D_1{}^{n_1 \times n_1} \times \cdots \times D_r{}^{n_r \times n_r}$. KG and all these matrix algebras are reversible. We have to prove that a reversible matrix algebra over a division algebra $D^{n \times n}$ is reduced which is to show $n = 1$. We assume $n \geq 2$ and focus on the matrices A and B which have only non-zero entries for $a_{21} = 1 = b_{22}$. Then $AB = 0$ and $BA = A \neq 0$ are valid.◇

Remark 15 Within the article [16] the authors classify reversible group algebras. In addition, they analyze on what terms the group algebra – also in positive characteristic – is symmetric. Symmetric group algebras are defined by the following rule: for all $a, b, c \in KG$ with $abc = 0$ the condition $acb = 0$ is valid.◇

Corollary 19 *Let K be a field and G a finite non-abelian hamiltonian group such that KG is semisimple. Then there exist a finite abelian group A of uneven order r and $n \in \mathbb{N}_0$ such that $G \cong Q_8 \times A \times (Z_2)^n$ is valid. Every Cartan subalgebra (= maximal torus) of $(KG)^\circ$ is of dimension $6 \cdot r \cdot 2^n$.*

Proof. We use some results which we will prove later on by character theory. All statements are a consequence of the corollaries 36, 33 and 29.◇

5.8.6 Reduced group algebras: the modular case

In chapter VII, lemma 1.10 of [26] it is proven that the factor algebra by the nilradical for a group algebra based on a finite group and a field of characteristic p is isomorphic to a direct products of matrix algebras over fields (of characteristic p). Within [62] the same result is proven by Benjamin Steinberg:

Lemma 10 *Let G be a finite group and K a field with $char(K) = p > 0$. The semisimple algebra $KG/rad(KG)$ is isomorphic to a direct sum of matrix algebras over field extensions of K.*

Add-on: If L is another field with $char(L) = p$ and F_p the field with p elements, then the following statement is valid:

$$LG/rad(LG) \cong KG/rad(KG) \cong K \otimes_{F_p} (F_pG/rad(F_pG)).$$

Proof. (Benjamin Steinberg) Let F_p be the field with p elements. $F_pG/rad(F_pG)$ is a separable algebra, because F_p is perfect. We want to prove the following statement:

$$KG/rad(KG) \cong K \otimes_{F_p} (F_pG/rad(F_pG)).$$

It is well-known that

$$K \otimes_{F_p} (F_pG/rad(F_pG)) = KG/(K \otimes_{F_p} rad(F_pG)).$$

is valid. $F_pG/rad(F_pG)$ is separable, and thus $K \otimes_{F_p} (F_pG/rad(F_pG))$ is semisimple. $K \otimes_{F_p} rad(F_pG)$ is a nilpotent ideal which is therefor identical to $rad(KG)$.

Finite division algebras are fields (Wedderburn, theorem 2) and F_p is perfect. Hence we derive

$$F_pG/rad(F_pG) \cong \prod_{i=1}^{m} M_{n_i}(L_i),$$

and L_i are finite separable field extensions of F_p. We obtain

$$KG/rad(KG) = K \otimes_{F_p} (F_pG/rad(F_pG)) = \prod_{i=1}^{m} M_{n_i}(K \otimes_{F_p} L_i).$$

As every L_i is a finite separable field extension, every $K \otimes_{F_p} L_i$ is a finite product of finite field extensions of K.

If L_i is a finite separable extension of F and K an extension of F, then we can use the theorem of the primitive element to conclude $L_i \cong F[t]/(p(t)F[t])$, and $p(t)$ splits into different linear factors in an algebraic closure F. We derive $K \otimes L_i \cong K[t]/(p(t)K[t])$, and $p(t)$ splits again in linear factors in an algebraic closure of K (containing F). The general case can be deducted by the result that a separable field extension is a direct limit of finite separable field extensions.◇

A consequence of this lemma is:

Theorem 41 *Let G be a finite group and K a field with $char(K) = p > 0$. KG is reduced if and only if KG is solvable. If KG is semisimple, then KG is reduced if and only if G is abelian.*◇

5.8.7 An example of Benjamin Steinberg

The following results and statements are based on a correspondence with Benjamin Steinberg which is not published yet. I want to say thank you to him for providing this example to me.

Let K be a field with $char(K) = 0$ and G a finite group. We focus on the monoid $(P(G); \cdot)$ based on the power set of G equipped with the complex product and the corresponding monoid algebra $KP(G)$. K is perfect, and by the theorem of Wedderburn-Malcev the nilradical of $KP(G)$ possesses an algebra complement. Benjamin Steinberg has proven that it is isomorphic to the direct product of all group algebras $K(G/N)$ for N being a normal subgroup of G. The monoid algebra is reduced if and only if $K(G/N)$ is reduced for every normal subgroup N of G. Every algebra $K(G/N)$ is an epimorphic image of KG (by using the linearization of G onto G/N). By usage of proposition 15 the algebra $KP(G)$ is reduced if and only if KG is reduced, and this statement is analyzed in this chapter. In particular, G is hamiltonian and the radical complements are isomorphic to the direct sum of KU for every subgroup = normal subgroup of G. Two open-ended questions are to calculate the maximal tori and Cartan subalgebras of $(KP(G))°$, also in the case that $KP(G)$ is not reduced.⋄

Section 5.8.

Reduced algebras

$\mathcal{S}$ maximal solvable subalgebras of A.
$\Rightarrow \mathcal{T}\otimes \text{rad}(A)$

$C_{\overline{\mathcal{T}\otimes \text{rad}(A)}}(\mathcal{T})$ Cartan subalgebras of A ↔ Cartan subalgebras of
$C_A(\mathcal{T})$ A/ρ

$\mathcal{T}$ maximal tori of A

{ woluble caft }
$\mathcal{S}$ solvable = reduced
$\text{rad}_A = 46/\text{rad}(4_0)$ is
a true woluble caft
always
a direction of matrix
algebras over fields!

{ reduced gorp algebras }
accepts G
+ characteristic thing
$\mathcal{J}_{46} = C_{(46)^0}$

diu Cartau
subalgebra
6.141.2a
(A odlcau,
141 werca

{ semisimple caft }
semisimple = separable

G haw Chaincau: $0 8 \times A \times \mathbb{Z}^2$?
and $9 \times 2 \times d 2 \times 1 = 0$ not separable
iff $(U_j$ H (Lal)) der wel primidit
root of uuity, $\partial l/14l$
$\Leftrightarrow 46 \times \text{wes.86} \Rightarrow 46$ symmetric
141 werca

5.8.8 Open-ended questions and exercises

Open-ended questions 7 *(i) Do other prominent families of reduced or solvable algebras exist and what are their Cartan subalgebras and maximal tori?*

(ii) How can we determine the Cartan subalgebras and maximal tori of a monoid algebra for an arbitrary monoid, for the power set monoid of a group or of a monoid? On what terms are these algebras reduced or solvable?

(iii) On what terms coincide the sets of maximal tori and Cartan subalgebras?

Excercise 190 *Let A be an associative finite-dimensional semisimple K-algebra. Prove that the conditions of being basic and reversible are equivalent. Is this true without assuming the semisimplicity of A?*

Excercise 191 *On what terms is the monoid algebra in example 5.8.7 solvable?*

Excercise 192 *Let K be a finite field. Prove the sum of two squares theorem: every element of the field is a sum of two squares. Does a special result exist for $char(K) = 2$? Is it possible to extend the two squares theorem to arbitrary fields of positive characteristic? Is the two squares theorem true for the fields $\mathbb{Q}, \mathbb{R}, \mathbb{C}$ and $\mathbb{H}$?*

Excercise 193 *In this exercise it is to be proven that we can derive from every associative right Artian K-algebra a basic algebra. In addition, by starting this procedure with a basic algebra the result is an isomorphic copy of the original algebra. The procedure is described within the tip. Thus, every reduced algebra can be derived by this procedure and the procedure is idempotent. (Tip: Use the results of section 2.4 in [39] and focus on the idempotents $e_1, \cdots, e_n$. Define the right ideals $R_i := e_i A$ for all $i \in \underline{n}$. Take a maximal set X of $\underline{n}$ such that all R_x are pairwise non-isomorphic and analyze the direct sum of these $R_x, x \in X$.)*

Excercise 194 *Is theorem 40 valid for an arbitrary semisimple associative algebra?*

Excercise 195 *Prove that subalgebras of reversible algebras are reversible, too. Is this statement true for factor algebras as well?*

Excercise 196 *Prove (if needed by using the literature quoted) lemma 9.*

Excercise 197 *Analyze the relations between reduced, reversible and symmetric (group) algebras! Use the articles cited in this section!*

Excercise 198 Let p be a prime number, K a field and G a finite non-abelian hamiltonian group such that KG is semisimple. Then there exist an abelian group A of uneven order r and $n \in \mathbb{N}_0$ such that $G \cong Q_8 \times A \times (Z_2)^n$ is valid. In the following cases decompose KG in simple algebras, decide whether KG is reduced or reversible and determine the dimension of the maximal tori and Cartan subalgebras:

(i) $n = 0$, $A = 1$, $K = \mathbb{Q}$

(ii) $n = 0$, $A = 1$, $K = \mathbb{R}$

(iii) $n = 0$, $A = 1$, $K = \mathbb{C}$

(iv) $n = 0$, $A = 1$, $K = \mathbb{Q}(i)$

(v) $n = 0$, $A = 1$, $K = \mathbb{Q}(\sqrt{2})$

(vi) K finite

(vii) K with $char(K) > 0$

(viii) $K = \mathbb{Q}(1 + i)$

(ix) $K = \mathbb{Q}(i + \sqrt{2})$

(x) $K = \mathbb{Q}(i + \sqrt{p})$

(xi) $K = \mathbb{Q}(1 + \sqrt{p}i)$

(xii) $K = \mathbb{Q}(i, \sqrt{2})$

(xiii) $K = \mathbb{Q}(i, \sqrt{p})$

(xiv) $K = \mathbb{Q}(i + 2^{\frac{2}{3}})$.

Excercise 199 Let A be a finite abelian group. Prove that the order a of A is determinable by $\sum_{d|a} |\{x \mid o(x) = d\}|$. (Tip: Lemma 9!) If A is cyclic, then determine each summand of the sum exactly by usage of Euler's totient function Φ.

Excercise 200 Is the tensor product of reduced associative algebras reduced, too? (Tip: $\mathbb{R}$ and $\mathbb{H}$)

Excercise 201 Let G be a finite group possessing exactly one minimal normal subgroup and K a field with $char(K) = 0$. Prove that G has a faithful irreducible representation. (Tip: Use the right regular representation!). Is the group G related to a division algebra? (Tip: Amitsur) Do you know an example for G? Use the results of this exercise for your example!

Excercise 202 True or false: A finite group whose order is the product of two (not necessary distinct) primes is abelian.

Excercise 203 Let G be a group and N, M normal subgroups of G. True or false:

(i) If N, M are nilpotent, then NM is nilpotent.

(ii) If N, M are abelian, then NM is abelian.

(iii) If N, M are simple, then NM is simple.

(iv) If N, M are solvable, then NM is solvable.

(v) If N, M are cyclic, then NM is cyclic.

Excercise 204 Let G be a group and N, M normal subgroups of G. True or false:

(i) If $G/N, G/M$ are nilpotent, then $G/(N \cap M)$ is nilpotent.

(ii) If $G/N, G/M$ are abelian, then $G/(N \cap M)$ is abelian.

(iii) If $G/N, G/M$ are simple, then $G/(N \cap M)$ is simple.

(iv) If $G/N, G/M$ are solvable, then $G/(N \cap M)$ is solvable.

(v) If $G/N, G/M$ are cyclic, then $G/(N \cap M)$ is cyclic.

Excercise 205 Let G be a finite group and N a normal subgroup of G. True or false: G/N and N' are nilpotent if and only if G is nilpotent. What is the answer for solvable groups? What is the answer by replacing N' with N for both questions? (Tip: characterizations of finite nilpotent groups.)

Excercise 206 Let D be a finite-dimensional division algebra and $n \in \mathbb{N}$. $D^{n \times n}$ possesses nilpotent elements different from zero if and only if $n \geq 2$ is valid.

Excercise 207 Are factor algebras of reduced algebras reduced, too?

Excercise 208 Prove remark 12!

Excercise 209 Prove remark 13!

Excercise 210 Let A, B be associative algebras. True or false: $A \times B$ is reduced if and only if A, B are reduced. On what terms is a semisimple associative algebra reduced or reversible?

Excercise 211 Let A, B be associative algebras. True or false: If $A \otimes B$ is reduced, then A, B are reduced.

Excercise 212 Let $n \in \mathbb{N}$ and A be an associative algebra. True or false: $A^{n \times n}$ is reduced if and only if A is reduced.

Excercise 213 Nilpotent and solvable algebras are reduced!

Excercise 214 Is a division algebra reduced? Is a field reduced?

Excercise 215 Are the algebras of lower and upper triangular matrices over a field reduced?

Excercise 216 Let K be a field. On what terms is the group algebra KS_3 reduced? In the reduced case determine the maximal tori and the Cartan subalgebras of $(KS_3)^\circ$.

Excercise 217 An associative finite-dimensional algebra over an algebraically closed field is reduced if and only if it is solvable.

Excercise 218 An associative finite-dimensional algebra over a finite field is reduced if and only if it is solvable. (Tip: theorem of Wedderburn for finite division algebras)

Excercise 219 An idempotent resp. nilpotent element of an associative algebra is nilpotent resp. idempotent if and only if it is the zero element.

Excercise 220 A nilpotent element of an associative algebra generates a nilpotent subalgebra.

Excercise 221 We focus on the power set $P(M)$ of a set M of order $\mid M \mid =: n$ and on a field K. Analyze the following statements:

(i) $(P(M); \cap)$ is an idempotent and commutative monoid.[8]

(ii) Is $\emptyset$ or M the neutral element?

(iii) Is (i) true for infinite sets?

(iv) What is the group of units of $(P(M); \cap)$?

(v) What are the answers for a group M?

(vi) The monoid algebra $KP(M)$ is isomorphic to K^{2^n}. (Tip: [76], chapter 1).

---

[8] An idempotent monoid is a monoid consisting of idempotent elements only.

141

Excercise 222 *We focus on the power set $P(M)$ of a set M of order $\mid M \mid$ $=: n$ and on a field K. Analyze the following statements:*

(i) $(P(M); \cup)$ is an idempotent and commutative monoid.

(ii) Is $\emptyset$ or M the neutral element?

(iii) Is (i) true for infinite sets?

(iv) What is the group of units of $(P(M); \cup)$?

(v) What are the answers for a group M?

(vi) The monoid algebra $KP(M)$ is isomorphic to K^{2^n}. (Tip: [76], chapter 1).

Tip: De Morgan laws and exercise 221.[9]

Excercise 223 *Define the symmetric and co-symmetric difference for sets. Determine for each pair of subsets of $\{1, 2, 3\}$ the symmetric and co-symmetric difference. If needed, then do a research in the literature for the corresponding definitions.*

Excercise 224 *We focus on the power set $P(M)$ of a set M of order $\mid M \mid$ $=: n$ and on a field K. Analyze the following statements:*

(i) $P(M)$ is with respect to the symmetric difference a commutative monoid.

(ii) Is $\emptyset$ or M the neutral element?

(iii) Is (i) true for infinite sets?

(iv) What is the group of units of $(P(M); \cup)$? Prove that the group of unit is elementary-2-abelian.

(v) What are the answers for a group M?

---

[9] Augustus De Morgan, born 27 June 1806, died 18 March 1871, was a British mathematician and logician. He formulated De Morgan's laws and introduced the term mathematical induction, making its idea rigorous. De Morgan was a brilliant and witty writer, whether as a controversialist or as a correspondent. In his time there flourished two Sir William Hamiltons who have often been conflated. One was Sir William Hamilton, 9th Baronet (that is, his title was inherited), a Scotsman, professor of logic and metaphysics at the University of Edinburgh; the other was a knight (that is, won the title), an Irishman, professor at astronomy in the University of Dublin. The baronet contributed to logic, especially the doctrine of the quantification of the predicate; the knight, whose full name was William Rowan Hamilton, contributed to mathematics, especially geometric algebra, and first described the Quaternions. De Morgan was interested in the work of both, and corresponded with both; but the correspondence with the Scotsman ended in a public controversy, whereas that with the Irishman was marked by friendship and terminated only by death.

(vi) The monoid algebra $KP(M)$ is isomorphic to K^{2^n}. (Tip: Solve part (iv) and analyze the group algebra of a direct product of groups.)

What are the answers for the so-called co-symmetric difference?

Excercise 225 *Which monoids and monoid algebras appearing in exercises 224, 221 and 222 are isomorphic? True or false: If two monoid algebras are isomorphic, then the underlying monoids are isomorphic, too.*

Excercise 226 *Study the following text ([90]):*

Carolyn Bean (reference below) proved a number of interesting group-theoretic results related to the symmetric difference operation, a few of which I will state here. Let X be a fixed set and let $P(X)$ be the set of all subsets of X. Bean proved that δ can be characterized as the unique group operation $*$ on $P(X)$ such that $A * B \subseteq A \cup B$ for all $A, B \in P(X)$. Define the co-symmetric difference operation δ_c on $P(X)$ by $A\delta_c B := X \setminus (A\delta B)$ for all $A, B \in P(X)$. Then one can show that $(P(X), \delta_c)$ is also an abelian group. Bean proved that δ_c can be characterized as the unique group operation $*$ on $P(X)$ such that $A \cap B \subseteq A * B$ for all $A, B \in P(X)$. Bean additionally proved that $(P(X), \delta, \cap)$ and $(P(X), \delta_c, \cup)$ are isomorphic commutative rings with identity. This can be found here: Carolyn Bean, Group operations on the power set, Journal of Undergraduate Mathematics 8.1, March 1976, Pages 13 to 17.

Excercise 227 *Let K be a field and G a finite group. We focus on the map from $P(G)$ into KG defined by $T \mapsto \overline{T}$. On what terms is T a subgroup or a normal subgroup? Is the function compatible with respect to the complex product? Describe the results for Q_8 and D_8 on their subgroups! on what terms is $\overline{T}$ central in KG?*

Excercise 228 *Use the results for reduced algebras in this section for solvable algebras! Do we have some additional information for solvable algebras as already proven in the previous sections?*

Excercise 229 *On what terms are the sets of maximal tori and Cartan subalgebras are identical for finite-dimensional associative unitary reduced algebras?*

Excercise 230 *(zero-extension) On what terms is the zero-extension (see exercise 39) a reduced algebra and how can we determine the maximal tori and Cartan subalgebras in this case with respect to the underlying algebra? Analyze the question by meaningful precondition for the underlying algebra!*

Excercise 231 *(eAe)* *On what terms is the algebra eAe (see exercise 136) a reduced algebra and how can we determine the maximal tori and Cartan subalgebras in this case with respect to the underlying algebra? Analyze the question by meaningful precondition for the underlying algebra or idempotent e!*

Excercise 232 *Let K be a field and G a finite non-abelian hamiltonian group. Determine on what terms KG is reduced and calculate the dimension of their maximal tori, Cartan subalgebras and its Lie nilradical!*

5.9 Associative algebras with separable factor algebra by its nilradical

5.9.1 A description by radical complements

Theorem 42 *Let A be an associative finite-dimensional unitary K-algebra with separable factor algebra by its nilradical and H a subset of A. The following statements are equivalent:*

(i) H is a Cartan subalgebra of A°.

(ii) A radical complement C of A and a Cartan subalgebra T of C° exists such that $H = C_A(T)$ is valid.

Add-on: The Cartan subalgebras of the radical complements are exactly their maximal tori which are exactly the maximal tori of A.

Proof. (i) $\to$ (ii): Let H be a Cartan subalgebra of A°. By using theorem 1 a maximal torus T of A exists such that $H = C_A(T)$ is valid. By 2.3.7 in [74] and lemma 4 T is contained in a radical complement C of A. It is straightforward to prove that T is a maximal torus of C, and by using theorem 36 the subalgebra T is a Cartan subalgebra of C°.

(ii) $\to$ (i): Let C be a radical complement of A and T a Cartan subalgebra of C°. By using theorem 36 the subalgebra T is a maximal torus of C and $T = C_C(T)$ is valid. The remark 13 ensures the identity $C_A(T) = C_C(T) \oplus C_{rad(A)}(T) = T \oplus C_{rad(A)}(T)$. T is a central radical complement in $C_A(T)$, and by using remark 10 we derive that $C_A(T)$ is Lie nilpotent. T is contained in a maximal torus S of A. Theorem 1 is used to derive that $C_A(S)$ is a Cartan subalgebra of A°. This subalgebra is maximal Lie nilpotent and contained in $C_A(T)$. By this we conclude that $C_A(T) = C_A(S)$ is valid, and we have proven that $C_A(T)$ is a Cartan subalgebra of A°.

Proof of the add-on: The first part is a direct consequence of theorem 36. Every maximal torus is a separable subalgebra, and by a general version of the theorem of Wedderburn-Malcev (see e.g. [74]) it can be conjugated into a radical complement. The second add-on can be derived performing this argumentation, too.⋄

5.9.2 A strategy for the determination of Cartan subalgebras

The results of theorems 36 and 42 can be used to describe a strategy for the determination of Cartan subalgebras of Lie algebras associated to associative unitary finite-dimensional algebras with separable factor algebra by their nilradical:

(1) Determine the maximal tori of the radical complements. This action is equivalent to determine the self-centralizing tori of the radical complements. For constructing one self-centralizing torus contained in a radical complement C, start with a (possibly great) torus T. Calculate the centralizer of this torus in C and search for a fully separable element t contained in this centralizer but not contained in T. The subalgebra $K[T,t]$ is again a torus containing T. Repeat this step until such an element t does not exist anymore.

(2) Calculate the centralizer in A of the tori constructed in (1) (see theorem 42). This centralizer is with respect to remark 13 and theorem 36 exactly $C_{rad(A)}(T) \oplus T$.⋄

We demonstrate this strategy: our first result is an alternative proof for the determination of Cartan subalgebras of solvable algebras as proven in theorem 24. Afterwards we determine some Cartan subalgebras for $(KD_{2n})°$.

5.9.3 Solvable algebras revised

Corollary 6 *Let A be a finite-dimensional associative unitary solvable K-algebra with a separable factor algebra by its nilradical. The Cartan subalgebras of $A°$ are exactly the centralizers of the radical complements of A.*

Proof. If C is a radical complement of A, then C is – by using lemma 1 – a torus. C is the only maximal torus of itself. The proof is finished by using theorems 36 and 42.⋄

5.9.4 Group algebras for dihedral groups

Let $n \in \mathbb{N}$, $G := D_{2n}$ the dihedral group of order $2n$ and K a field with $char(K) = p$. Then $a, b \in G$ exist with $o(a) = n$, $o(b) = 2$, $G = \langle a, b \rangle$ and $a^b = a^{-1}$. $G = \{1, a, ..., a^{n-1}, b, ab, ..., a^{n-1}b\}$ is valid. If $p > 0$ is not a divisor of n or $p = 0$ is valid, then – using a theorem of Maschke – $K\langle a \rangle$ is semisimple and commutative. By 1.9.4.2 in [74] the subalgebra $K\langle a\rangle$ is separable over K, and by lemma 4 we conclude that $K\langle a\rangle$ is a torus of KG. We analyze the centralizer of $K\langle a\rangle$ in KG. $KG = K\langle a\rangle \oplus \langle b, ab, ..., a^{n-1}b\rangle_K$ is valid, and $K\langle a\rangle$ centralizes $K\langle a\rangle$. Let $x \in \langle b, ab, ..., a^{n-1}b\rangle_K$ represented as $x = \sum_{i=0}^{n-1} k_i a^i b$. x centralizes $K\langle a\rangle$ if and only if $\sum_{i=0}^{n-1} k_i a^i b^a = \sum_{i=0}^{n-1} k_i a^i b$ is valid. Because of $b^a = a^{-2}b$ this is the case if and only if $a^{-2}(\sum_{i=0}^{n-1} k_i a^i) = \sum_{i=0}^{n-1} k_i a^i$ is true. If 2 is not dividing n (which is equivalent to $\langle a^{-2}\rangle = \langle a\rangle$), then x centralizes $K\langle a\rangle$ if and only if $K\langle a\rangle$ acts trivial on $\sum_{i=0}^{n-1} k_i a^i \in K\langle a\rangle$. This statement is equivalent to $\sum_{i=0}^{n-1} k_i a^i \in \langle \sum_{i=0}^{n-1} a^i \rangle_K$. We derive $C_{KG}(K\langle a\rangle) = K\langle a\rangle \oplus \langle b + ab + ... + a^{n-1}b\rangle_K = K\langle a\rangle \oplus \langle \sum_{g \in G} g\rangle_K$.

If p is dividing the order of G – which is by our assumptions only the case for $p = 2$ –, then the one-dimensional K-subspace $\langle \sum_{g \in G} g\rangle_K$ is a zero-ideal and thus contained in the radical of KG. Let $KG/rad(KG)$ be separable. The the torus $K\langle a\rangle$ is by using lemma 4 and 2.3.7 in [74] contained in a in radical complement C of KG. Remark 13 ensures the identity $C_{KG}(K\langle a\rangle) = C_{rad(KG)}(K\langle a\rangle) \oplus C_C(K\langle a\rangle)$. In addition, $C_{KG}(K\langle a\rangle) = K\langle a\rangle \oplus \langle \sum_{g \in G} g\rangle_K$, $K\langle a\rangle \subseteq C_C(K\langle a\rangle)$ and $\langle \sum_{g \in G} g\rangle_K \subseteq C_{rad(KG)}(K\langle a\rangle)$ are valid. We conclude that $K\langle a\rangle$ self-centralizing in C. The theorems 36 and 42 provide that $C_{KG}(K\langle a\rangle) = K\langle a\rangle \oplus \langle \sum_{g \in G} g\rangle_K$ is a $(n+1)$-dimensional Cartan subalgebra of $(KG)^\circ$.

Let p not dividing the order of G. The element $s := \sum_{g \in G} g$ is diagonalizable with respect to $s^2 = |G| \, s$. Hence it is separable over K. We use our strategy to obtain that $C_{KG}(K\langle a\rangle)$ is a torus of KG which is obviously self-centralizing in KG. Hence $C_{KG}(K\langle a\rangle)$ is a Cartan subalgebra of $(KG)^\circ$ also in this case.⋄

Section 5.9

> Associative algebras with separate
> tropical algebra & their tropicalizations
> → Separate case
>   in an easy
>   context

Example: for $(k*\Omega_n)^o$

[Diagram: parallelogram with vertices labeled σ, B, T_{max}, $G(T_{max})$, A]

- A — Cartan subalgebra of A
- T_{max} — Tropical component
         = maximal tori of T
         = maximal tori of A
         = Cartan subalgebra of A
- $G(T_{max})$ ⊆ ∈ — Cartan subalgebra of A
- B ⊆ ∈ — Cartan subalgebra of T

J_A maximal tori of A

$\|$

J_T maximal tori of $T \xrightarrow{\text{trop}(\cdot)} \mathcal{T}$ ≅ Tropicale

5.9.5 Open-ended questions and exercises

Open-ended questions 8 *(i) For a field K and a finite group G determine all maximal torus and all Cartan subalgebras of the associated Lie algebra of KG!*

(ii) On what terms are the sets of maximal tori and Cartan subalgebras identically?

Excercise 233 *Use the argumentation of the section 5.9.4 about group algebras of dihedral groups and extend it for the case $2 \mid n$!*

Excercise 234 *Use the argumentation of the section 5.9.4 about group algebras of dihedral groups and extend it to group algebras of quaternion groups Q_{4n}!*

Excercise 235 *An associative algebra is called local if and only if it possesses exactly one maximal ideal. Prove that local algebras are reduced and its factor algebra by its radical is isomorphic to a division algebra.*

Excercise 236 *Study within [85] all statements for local algebras and prove them (possibly by usage of additional literature)!*

Excercise 237 *Let K be a field and G a finite group. On what terms is the group algebra KG local? Determine all Cartan subalgebras of $(KG)^\circ$ for a local group algebra KG. (Tip: see [75])*

Excercise 238 *Prove that a module is indecomposable if and only if the algebra of all endomorphism of the module is local.*

Excercise 239 *True or false: Direct sums of reduced algebras are reduced. Direct sums of local algebras are local. Direct sums of local algebras are reduced. Every reduced algebra is a direct sum of local algebras.*

Excercise 240 *Let A be a finite-dimensional associative unitary local algebra with a separable factor algebra by its nilradical. In what way is it possible to determine the Cartan subalgebras and maximal tori of A°? Use the results for reduced algebras of this chapter, the strategy 5.9.2 and the main theorem for division algebras.*

Excercise 241 *For this exercise let $K := \mathbb{Q}$.*

(i) Determine a maximal torus and a Cartan subalgebra of $(KQ_8)^\circ$ and calculate their dimension! In what way do the results change if K is replaced by the real or complex number field? Is it possible to determine all maximal tori and Cartan subalgebras of $(KQ_8)^\circ$? Is the algebra $(KQ_8)^\circ$ reduced or local?

(ii) Analyze part (i) if the algebra $(KQ_8)^\circ$ is replaced by the lower triangular matrices of $(K^{8\times 8})^\circ$!

(iii) Analyze part (i) for the direct sum of $(KQ_8)^\circ$ and the lower triangular matrices of $(K^{8\times 8})^\circ$

Excercise 242 *(zero-extension) We focus again on exercise 39. Analyze the determination of the Cartan subalgebras and maximal tori of the Lie algebra associated to the zero-extension of an associative algebra possessing a separable factor algebra by its nilradical. If the analysis is to complex, then assume e.g. that the original algebra is separable.*

Excercise 243 *(eAe) Let A be an associative unitary finite-dimensional algebra with separable factor algebra by its nilradical and e an idempotent of A. Analyze the determination of the Cartan subalgebras and maximal tori of the Lie algebra associated to eAe. If the analysis is to complex, then assume e.g. that the idempotent is central.*

5.10 Natural determination of Cartan subalgebras

In the previous sections we have determined the Cartan subalgebras for diverse classes of algebras: solvable, division algebras, simple algebras, semisimple algebras, reduced algebras etc. For direct products the determination was purely natural: its direct product of the Cartan subalgebra of the factors. In this section we analyze this question further: in what way is the determination of the Cartan subalgebras of special constructions of algebras to be answered naturally with respect to the construction?

5.10.1 Sub- and factor structures

Example 5 In this example we focus on the determination of Cartan subalgebras for subalgebras, left and right ideals. A natural question is whether they can be calculated by the intersection of a Cartan subalgebra of the underlying algebra with its substructure. But the answer is negative in general: let $n \in \mathbb{N}$, K be a field and $A := K\Pi_n$ the Solomon-Tits algebra (see e.g. [76]). In [76] it is proven that the radical complements are exactly the Cartan subalgebras of A°. They have trivial intersection with the nilradical of A which is Lie nilpotent. The unital subalgebra $rad(A) \oplus K \cdot 1$ is Lie nilpotent, too, but the intersection with every Cartan subalgebra is one-dimensional.◇

Proposition 19 *Let A be a K-algebra, C a Cartan subalgebra of A° and I an ideal of A contained in C. Then $((C+I)/I)^\circ$ is a Cartan subalgebra of $(A/I)^\circ$.*

Proof. By using proposition 8 the Lie algebra C is a subalgebra of A. It is well-known that $(C+I)/I$ and $C/(C \cap I)$ are isomorphic. This factor algebra is Lie nilpotent because C is Lie nilpotent, too. We have to prove that the factor algebra is self-normalizing as a Lie algebra. Let $x \in A$ with $(x+I) \circ ((C+I)/I) \subseteq (C+I)/I$. For all $c \in C$ the identity $(x \circ c) + I \in (C+I)/I = C/I$ is valid, and this implies $x \circ c \in C + I = C$. C is self-normalizing and the proof is finished. ⋄

Definition and remark 5 *(Engel subalgebras)* Let V be a vector space of finite dimension over a field K and f an endomorphism of V. By $V_0(f)$ we denote the Fitting null-component of V with resp. to f. If $\mathcal{L}$ is a subset of $End_K(V)$, then $V_0(\mathcal{L})$ is the Fitting null-component V with resp. to $\mathcal{L}$, and its defined by:

$$V_0(\mathcal{L}) = \{v \in V \,|\, \forall f \in \mathcal{L} \,\exists n \in \mathbb{N}:\, vf^n = 0\}.$$

Let L be a finite-dimensional Lie algebra. Then it is well-known that a nilpotent subalgebra H of L is a Cartan subalgebra if and only if $H = L_0(\text{ad}\,H)$ is valid (see e.g. [29], proposition 1, chapter III,1).

In addition, if the condition $dim_K(L) < |\,K\,|$ is valid, then the Cartan subalgebras are exactly the nilpotent resp. minimal Engel[10] subalgebras. These can be presented as $L_0(ad(x))$ for an element $x \in L$. Engel subalgebras have a nice feature: every subalgebra in which they are contained is self-normalizing. In particular, Engel subalgebras are self-normalizing (see e.g. [39]).⋄

By using Engel subalgebras we are able to determine Cartan subalgebras of factor algebras in a natural manner. Every factor algebra induces an epimorphism from the underlying algebra onto its factor algebras.

---

[10]Friedrich Engel (born December 26, 1861, died September 29, 1941) was a German mathematician. Engel was born in Lugau, Saxony, as the son of a Lutheran pastor. He attended the Universities of both Leipzig and Berlin, before receiving his doctorate from Leipzig in 1883. Engel studied under Felix Klein at Leipzig, and collaborated with Sophus Lie for much of his life. He worked at Leipzig (from 1885 to 1904), Greifswald (from 1904 to 1913), and Giessen (from 1913 to 1931). He died in Giessen. Engel was the co-author, with Sophus Lie, of the three volume work Theorie der Transformationsgruppen (publ. from 1888 to 1893; tr., 'Theory of transformation groups'). Engel was the editor of the collected works of Sophus Lie with 6 volumes published between 1922 and 1937; the seventh and final volume was prepared for publication but appeared almost twenty years after Engel's death. He was also the editor of the collected works of Hermann Grassmann. Engel translated the works of Nikolai Lobachevski from Russian into German, thus making these works more accessible. With Paul Stäckel he wrote a history of non-Euclidean geometry (Theorie der Parallellinien von Euklid bis auf Gauss, 1895). With his former student Karl Faber, he wrote a book on the theory of partial differential equations of the first order using methods of Lie group theory. In 1910 Engel was the president of the Deutsche Mathematiker-Vereinigung.

Corollary 20 *Let L, M finite-dimensional K-Lie algebras, φ an epimorphism from L onto M, C a Cartan subalgebra of L and $dim_K(L) < \mid K \mid$. Then $C\varphi$ is a Cartan subalgebra of M.*

In particular, if I is an ideal of L, then $(C + I)/I$ is a Cartan subalgebra of L/I.

Proof. We use the results included in definition and remark 5. The Cartan subalgebra C is a minimal Engel subalgebra, and every subalgebra containing C is self-normalizing. In particular, $C + Kern\varphi$ is self-normalizing. As an epimorphic image $C\varphi$ is nilpotent. Let z an element of the normalizer of $C\varphi$ in M. Then there exists an element $x \in A$ with $z = x\varphi$. For all $c \in C$ the condition $(x\varphi)(c\varphi) \in C\varphi$ is valid. Thus for all $c \in C$ the identity $(xc)\varphi \in C\varphi$ is true. Hence for all $c \in C$ an element $c\prime \in C$ exists such that $(xc)\varphi = c\prime\varphi$ is valid. This statement is equivalent to the condition that for all $c \in C$ an element $c\prime \in C$ exists such that $(xc - c\prime)\varphi = 0$ is valid. We conclude for all $c \in C$ that the condition $xc \in C + Kern\varphi$ is valid. It is straightforward to verify that for all $c \in C, t \in Kern\varphi$ the element $x(c+k)$ is contained in $C + Kern\varphi$. We have proven that x normalizes the subalgebra $C + Kern\varphi$, and this subalgebra is self-normalizing. We conclude that $x \in C + Kern\varphi$ and therefor $x\varphi \in C\varphi$ are valid.⋄

The determination of all Cartan subalgebras is not known by the author.

5.10.2 Direct products

Within theorem 35 we have determined the Cartan subalgebras of Lie algebras associated to direct products of associative algebras. The answer was purely natural, because all of them can be created by the Cartan subalgebras of the factors.⋄

5.10.3 Adjunction of a unit

Within proposition 13 we have proven the following natural result: if (A, K) is the adjunction of a unit with respect to the K-algebra A and C is a Cartan subalgebra of the Lie algebra A°, then $(C, K)^\circ$ is a Cartan subalgebra of the Lie algebra $(A, K)^\circ$.

We prove now that all Cartan subalgebras can be determined by this procedure. For proving this, we use the some basic results on the adjunction of a unit included in volume II of this series (lemma3, section 6.4 in [78]). Every Cartan subalgebra is a unital associative subalgebra. By the quoted lemma all unital subalgebras of (A, K) are of the form (K, C) for a subalgebra C of A. (K, C) is Lie nilpotent and hence C is Lie nilpotent, too. In addition, $N_{(K,A)^\circ}((K, C)) = (K, N_{A^\circ}(C))$ is valid, and thus C is a Cartan subalgebra of A°.⋄

5.10.4 Cyclic algebras

For a cyclic algebra based on a Galois extension$(K;L)$ we have demonstrated within example 4 that L is a Cartan subalgebra of the Lie algebra associated to the cyclic associative algebra. The determination of all Cartan subalgebras is not known by the author.⋄

5.10.5 Tensor products and extension of the base field

We have used within several proofs that for the extension of the base field $A \otimes L$ of an algebra A the extension of the base field of a Cartan subalgebra is a Cartan subalgebra, too (see e.g. [29]).

Proposition 20 *Let A, B finite-dimensional associative unitary K-algebras and C resp. D an unital K-subalgebra of A resp. B. The identity*

$$N_{(A \otimes B)^\circ}(C \otimes D) = N_{A^\circ}(C) \otimes N_{B^\circ}(D)$$

is valid.

Proof. The proof is to be done by the reader (see exercise 247).⋄

Within corollary 2 we have proven by some special assumptions that Lie nilpotency and tensor products are compatible. By this and the previous propositions 20 and 8 we derive the following natural result:

Theorem 43 *Let A, B be finite-dimensional associative unitary K-algebras with separable factor algebra by their nilradical and C resp. D a Cartan subalgebra of A° resp. B°. Then $C \otimes D$ is a Cartan subalgebra of $(A \otimes B)^\circ$.*⋄

The construction of all Cartan subalgebras of $(A \otimes B)^\circ$ is not known to the author.

5.10.6 Matrix algebras

Remark 16 Let A be a K-algebra and $n \in \mathbb{N}$. A natural questions is whether $C^{n \times n}$ – for a Cartan subalgebra C of A° – is a Cartan subalgebra of $(A^{n \times n})^\circ$. But by using remark 9 we conclude that the algebra $C^{n \times n}$ is for all $n \geq 2$ not Lie nilpotent.

Within several proofs we have used that – for a field K and a natural number n – the set of diagonal matrices is a Cartan subalgebra of $gl(n, K)$. In addition, all Cartan subalgebras are conjugated if K is algebraically closed. We generalize this result to $(A^{n \times n})^\circ$.⋄

Proposition 21 *Let A be an associative unitary K-algebra, $n \in \mathbb{N}$ and C a Cartan subalgebra of A°. Then the algebra of diagonal matrices over C – which are denoted by $D(n, C)$ – is a Cartan subalgebra of $(A^{n \times n})^\circ$.*

Proof. C is with respect to proposition 8 a subalgebra of A. As a consequence $D(n,C)$ is a Lie subalgebra of $(A^{n\times n})^\circ$.

Let X a matrix of the Lie normalizer of $D(n,C)$ in $(A^{n\times n})^\circ$. By using the special diagonal matrices possessing only one entry different to zero on the diagonal – and this entry is exactly 1 – and the definition of the Lie normalizer we prove that X is a diagonal matrix over A. Let $c \in C$. Then X normalizes the diagonal matrix possessing n-times the entry c on the diagonal. Therefor for all $i \in \underline{n}$ the element $X_{ii} \circ c$ is contained in C. Hence X_{ii} normalizes all elements of C, and X is an element of $D(n,C)$.

The Lie nilpotency of $D(n,C)$ can be derived by the correspondent one for C: a n-fold Lie product of elements of $D(n,C)$ is a diagonal matrix over C possessing on the diagonal values which are n-fold Lie products with elements of C.⋄

The construction of all Cartan subalgebras of $(A^{n\times n})^\circ$ is not known to the author. Matrix algebras are related to tensor products by the identity $A^{n\times n} \cong K^{n\times n} \otimes A$.

5.10.7 Group algebras

The following question will be analyzed in part II of this volume which might be a natural question as well: let K be a field, G a solvable finite group and C a Carter subgroup of G. Is KC a Cartan subalgebra of $(KG)^\circ$? We will prove that the algebra KC is Lie nilpotent but not self-normalizing.

There will be several results with resp. to Cartan subalgebras of group algebras presented in this series. For solvable group algebras there is a natural determination possible (see 5.5.3). A general determination of all Cartan subalgebras is not known by the author.⋄

Section 10.

Compliance with constructions of adjoints?
A - Algebra - ?
C - CTA - O

- adjunctive ?/unit
  A4 with C4

- cyclic algebra, strict woowwap subfield

- direct product
  A×B with CTA C×D

- factor algebra
  A/I with CTA C+I/I (I ...)
  (inj? substitution)

- tensor product A⊗B
  with CTA C⊗D

- matrix algebra A^n
  with CTA JC(C_n)
  $\begin{pmatrix} C_1 & 0 \\ 0 & C_n \end{pmatrix}$

- no compliance with substructure

5.10.8 Open-ended questions and exercises

Open-ended questions 9 *(i) Is it possible to determine all Cartan subalgebras for the algebra constructions presented in this section?*

(ii) Construct all Cartan subalgebras for factor algebras in the case $dim_K(L) \geq |K|$.

(iii) Determine and construct all Cartan subalgebras for group algebras.

(iv) Determine and construct all Cartan subalgebras for tensor products.

(v) On what terms are Cartan subalgebras and maximal tori identical?

Excercise 244 *By using the results of section 5.10.5 prove again proposition 21. (Tip: $A^{n \times n} \cong K^{n \times n} \otimes A$)*

Excercise 245 *Let A be a real quaternion algebra. True or false: A is isomorphic to its inverse algebra. Is this statement true or false for an arbitrary quaternion algebra? Determine the Cartan subalgebras of $A \otimes A$. (Tips: $A \otimes A$ is isomorphic to $\mathbb{C}^{2 \times 2}$; results of section 5.10.5)*

Excercise 246 *Every homomorphic image of a nilpotent Lie algebra is nilpotent, too. Is this statement true for abelian and solvable Lie algebras as well?*

Excercise 247 *Prove proposition 20!*

Excercise 248 *Let A be the algebra of lower triangular matrices (related to an arbitrary field K and a natural number n), D the set of diagonal matrices and J the nilradical of A. True or false: For all $i \in \underline{n}$ the set $(D + J^i)/J^i$ is a Cartan subalgebra of $(A/J^i)^\circ$? In what way could the size of the field be relevant for this problem?*

Excercise 249 *Let K be a field, G a finite group and U a subgroup of G. True or false: The dimension of a Cartan subalgebra of $(KU)^\circ$ is not greater than the one for $(KG)^\circ$? Every Cartan subalgebra of $(KU)^\circ$ the intersection of a Cartan subalgebra of $(KG)^\circ$ with KU. Every Cartan subalgebra of $(KU)^\circ$ can be extended to one of $(KG)^\circ$.*

Excercise 250 *(zero-extension) Transfer exercise 39 to the topic of Cartan subalgebras! Analyze the transfer of a maximal torus in a radical complement of the zero-extension and begin with the separable case of A.*

Excercise 251 *(eAe) Transfer the exercise 136 to the topic of Cartan subalgebras! Analyze the transfer of a maximal torus in a radical complement to eAe and begin with the separable case of A.*

Excercise 252 *Let L be a finite-dimensional Lie algebra. L is nilpotent if and only if the intersection of $nil(L)$ with every subalgebra U of L is exactly $nil(U)$.*

Excercise 253 *Let L be a finite-dimensional Lie algebra and C a Cartan subalgebra of L. L is nilpotent if and only if the intersection of C with every subalgebra U of L as a Cartan subalgebra of U.*

Chapter 6

Summation formulas for the dimension of maximal tori in group algebras

In this chapter we derive some formulas for the dimension of the maximal tori in group algebras. For semisimple group algebras maximal tori and Cartan subalgebras are identical. The general summation formulas are represented in the style of the article [59]. The author thanks Salvatore Siciliano for his support. The general summation formulas are applied to several group classes and are used to bound the dimension. We are using several facts of the character theory of finite groups. An introduction to this theory can be found e.g. in the benchmark of Martin Isaacs [27].

6.1 The semisimple case

We are focussing on the semisimple case of group algebras. Salvatore Siciliano proved that the dimension of every Cartan subalgebra of $(KG)^\circ$ is the sum of the degrees of the complex irreducible characters of G. For a representation resp. character α we denote by $deg(\alpha)$ or $grad(\alpha)$ its degree which is the dimension of the corresponding KG-module.

Theorem 44 *(First summation formula by Salvatore Siciliano) Let G be a finite group, K a field such that KG is semisimple and $\chi_1, \chi_2, \cdots, \chi_{k(G)}$ the complex irreducible characters of G. The following statement are valid:*

(i) *The Cartan subalgebras of $(KG)^\circ$ are exactly the maximal tori of KG.*

(ii) *Every Cartan subalgebra H of $(KG)^\circ$ possesses the dimension*

$$\dim_K H = \sum_{i=1}^{k(G)} \deg(\chi_i).$$

Proof. By using example 1.9.4.2 in [74] we derive that KG is a separable algebra. By theorem 36 the first part is proven. Let F be an algebraically closed superfield of K. If H is a K-Cartan subalgebra of $(KG)^\circ$, then – by using [29] – the F-algebra $H \otimes_K F$ is a Cartan subalgebra of the F-Algebra $(KG \otimes_K F)^\circ$. This tensor product is isomorphic as F-algebra to FG (see exercise 300). It is well-known (see e.g. [49]) that the dimension formula $dim_F(H \otimes_K F) = dim_K(H)$ is valid. Therefor we can assume that K is algebraically closed. By the theorem of Wedderburn-Artin KG is isomorphic to a direct sum of matrix-algebras $KG \cong K^{n_1 \times n_1} \oplus K^{n_2 \times n_2} \oplus \cdots \oplus K^{n_{k(G)} \times n_{k(G)}}$. From theorem 36 we derive that H is isomorphic to a direct sum of maximal tori related to the matrix-algebra decomposition. We have already used the result in [29] that every maximal torus is isomorphic to the subalgebra of diagonal matrices. Thus the K-Dimension of H is exactly $\sum_{i=1}^{k(G)} n_i$. Each of the values n_i is a degree of a irreducible KG-representation. We finish the proof by theorem 12.11 in [25], Chapter 5: only the complex numbers are relevant.⋄

Let K be a field and G a finite group such that KG is semisimple. By $cta((KG)^\circ)$ we denote the dimension of all maximal tori = Cartan subalgebras of $(KG)^\circ$.

In the next parts we use theorem 44 and some basic and new results of the character theory of finite groups to calculate this sum for some special classes of groups and to bound this sum, too.

6.2 Upper and lower bounds

Corollary 21 *(square root of group order) Let G be a finite K a field such that KG is semisimple. The inequality*

$$\sqrt{|G|} \leq cta((KG)^\circ) \leq |G|$$

is valid.

Proof. The squares of the degrees of all complex irreducible characters is the order of G. We finish the proof by using theorem 44. ⋄

Corollary 22 *(subgroup-bound) Let G be a finite group, U a subgroup of G, A a (preferably maximal) abelian subgroup of G and K a field such that KG is semisimple. The inequality*

$$max\{|A|, cta((KU)^\circ)\} \leq cta((KG)^\circ)$$

is valid.

Proof. By using a theorem of Maschke and the theorem of Lagrange we derive that KU is semisimple, too. Every maximal torus of KU is a torus of KG, too. KA is a torus itself. We finish the proof by using theorem 44. ⋄

The next three inequalities are based on results and remarks of Geoffrey Robinson (in mathoverflow.net) for which I want to say thank you to him. For a finite group G we denote by $t(G)$ the number of involutions (= elements of order 2) of G.

Corollary 23 *(involution-bound) Let G be a finite group and K a field such that KG is semisimple. The inequality*

$$cta((KG)^\circ) \geq 1 + t(G)$$

is valid.

Proof. By using a theorem of Frobenius-Schur we conclude that for $n \in \mathbb{N}$ the nth root number function is a class function and hence a linear combination of irreducible complex characters. In the special case $n = 2$ only the values 1, 0 and -1 are appearing in this combination. The value of the second root number function at 1 is exactly $1 + t(G)$. Hence the sum of degrees is at least this value (see e.g. [27], pages 49ff. for results for the nth root number functions). We finish the proof by using theorem 44. ⋄

Corollary 24 *(Robinson, 2014) Let G be a finite non-abelian group, K a field such that KG is semisimple and $\chi_k(1)$ the maximal dimension of the irreducible complex characters of G. The inequality*

$$cta((KG)^\circ) \leq \sqrt{k(G) + 1 - \frac{|G|}{\chi_k(1)}}$$

is valid.

Proof. The upper bound is presented in the following articles of Geoffrey Robinson: [53], [54] and [55]. We finish the proof by using theorem 44. ⋄

Corollary 25 *(Robinson, 2014) Let G be a finite nilpotent group, K a field such that KG is semisimple and $\chi_k(1)$ the maximal dimension of the irreducible complex characters of G. The inequality*

$$cta((KG)^\circ) \leq \sqrt{(k(G) - 1) |G| + (\chi_k(1))^2}$$

is valid.

Proof. The upper bound is presented in the following articles of Geoffrey Robinson: [53], [54] and [55]. We finish the proof by using theorem 44. ⋄

Corollary 26 *(meta-abelian groups)* Let G be a meta-abelian finite group and K a field such that KG is semisimple. Let N be a normal subgroup containing the derived subgroup of G (and preferable maximal with this condition). The inequality

$$cta((KG)^\circ) \leq |G/G'| + (k(G) - |G/G'|) \cdot |G/N|$$

is valid.

Proof. There are $|G/G'|$ linear characters. The degrees of all the other characters are bounded by a theorem of Ito by the index of N in G. We finish the proof by using theorem 44. ◇

In the following corollary the classification of simple groups is used by Berkovitch and Mann:

Corollary 27 *(Berkovitch and Mann, 1998)* Let G be a finite non-solvable group and K a field such that KG is semisimple. Then the strong inequality

$$cta((KG)^\circ) > 2 \cdot |G/G'|$$

is valid.

Proof. Berkovitch and Mann proved in their article in [8] that the sum of all complex irreducible characters has the lower bound as presented. We finish the proof by using theorem 44. ◇

Corollary 28 *(Heffernan and MacHale, 2008)* Let G be a finite group, p a (preferably great) prime divisor of the order of G, $n \in \mathbb{N}$, $p^n \mid\mid |G|$ and K a field such that KG is semisimple. Then the following inequality is valid:

$$cta((KG)^\circ) \geq p^{\lfloor \frac{\lfloor\sqrt{8n+1}\rfloor - 1}{2} \rfloor}.$$

Proof. Heffernan and MacHale proved in their article [18] that the sum of all complex irreducible characters has the lower bound as presented. We finish the proof by using theorem 44. ◇

6.3 Special classes of groups

Corollary 29 *(abelian groups)* Let G be a finite group and K a field such that KG is semisimple. G is abelian if and only if the dimension of all Cartan subalgebras of $(KG)^\circ$ is exactly $|G|$. This is equivalent to the statement that KG is Lie nilpotent.

Proof. It is well-known that G is abelian if and only if every complex irreducible character is linear. The sum of degrees of all complex irreducible characters is identical to the sum of squares of these degrees (and this sum is $\mid G\mid$) if and only if every degree is 1 (because the degrees are natural numbers). Now the proof is finished by using theorem 44 and the fact that Cartan subalgebras are Lie nilpotent. ⋄

Corollary 30 *(pq-groups) Let p, q be prime numbers, G a finite group of order pq and K a field such that KG is semisimple. G is abelian or $p \mid (q-1)$ is valid. In this cases $cta((KG)^\circ) = pq$ resp. $cta((KG)^\circ) = q + p - 1$ are valid.*

Proof. By using the classification of pq-groups in [44] G is abelian or p is a divisor of $q - 1$. In the latter case we conclude by using theorem 25.10 in [44] that there are q linear and $\frac{p-1}{q}$ q-dimensional complex irreducible characters. The sum of their degrees is $q + p - 1$. The proof is finished by using corollary 29 and theorem 44. ⋄

Corollary 31 *(extra-special p-groups) Let K be a field, p a prime number, $char(K) \neq p$, $n \in \mathbb{N}$ and G an extra-special p-group of order p^{2n+1}. Then the identity*

$$cta((KG)^\circ) = p^n(p^n + p - 1)$$

is valid.

Proof. By using the article [79] we conclude that there are p^{2n} linear and $(p-1)$ p-dimensional irreducible complex characters of G. Theorem 44 lets us derive that the dimension of every Cartan subalgebra of $(KG)^\circ$ is exactly $p^{2n} + (p-1)p^n$, and this finish the proof. ⋄

Corollary 32 *(dihedral groups) Let $n \in \mathbb{N}$, $n \geq 3$, $G = D_{2n}$ the dihedral group of order $2n$ and K a field with $char(K) \neq 2$ and no divisor of n. If n is even resp. uneven, then the identities $cta((KG)^\circ) = n + 2$ resp. $cta((KG)^\circ) = n + 1$ are valid. In addition, all even numbers not less than 4 are dimension of Cartan subalgebras for group algebras of non-abelian groups.*

Proof. Let $a, b \in G$, $o(a) = n$ and $o(b) = 2$ such that G is generated by a, b and $a^b = a^{-1}$ is valid. The commutator of a, b is exactly a^{-2}. If n is uneven, then the derived subgroup is of index 2. In the other case its index is 4. Hence we have 2 resp. 4 linear complex characters within these cases. The conjugacy classes of G are b^G, $(ab)^G$, $\{1\}$ and for n being even resp. uneven are $\{a^{\frac{n}{2}}\}$ and $\{a^r, a^{-r}\}$ for $r \in \frac{n}{2} - 1$ resp. $\{a^r, a^{-r}\}$ for $r \in \frac{n-1}{2}$. If n is uneven, then b^G, $(ab)^G$ are within one class. Therefor the class number of G is $\frac{n}{2}+3$ resp. $\frac{n-1}{2}+2$. a is generating a cyclic normal subgroup of index 2. By using a theorem of Ito all non-linear complex irreducible characters are of degree 2. We conclude with theorem 44 that the dimension of all Cartan subalgebras of $(KG)^\circ$ is exactly $4 \cdot 1 + (\frac{n}{2} - 1 + 4 - 4) \cdot 2 = n+2$ resp. $2 \cdot 1 + (\frac{n-1}{2} + 2 - 2) \cdot 2 = n+1$. ⋄

Corollary 33 *(quaternion groups) Let $n \in \mathbb{N}$, $n \geq 2$, $G = Q_{4n}$ the quaternion group of order $4n$ and K a field with $char(K) \neq 2$ and no divisor of n. The identity*

$$cta((KG)^\circ) = 2n+2$$

is valid.

Proof. Let $a, b \in G$, $o(a) = 2n$ and $o(b) = 4$ such that G is generated by a, b and $a^b = a^{-1} b^2$ is valid. The derived subgroup is generated by a^2. Hence there are 4 linear irreducible complex characters. a is generating a cyclic normal subgroup of index 2. By using a theorem of Ito all non-linear complex irreducible characters are of degree 2. The class number of G is $k(G) = n+3$. We conclude with theorem 44 that the dimension of all Cartan subalgebras of $(KG)^\circ$ is exactly $4 \cdot 1 + (n-1) \cdot 2 = 2n+2$. ⋄

Corollary 34 *(semi-dihedral groups) Let $n \in \mathbb{N}$, $G = SD_{2^n}$ the semi-dihedral group of order 2^n and K a field with $char(K) \neq 2$. The identity*

$$cta((KSD_{2^n})^\circ) = 2^{n-1} + 2 = cta((KQ_{2^n})^\circ) = cta((KD_{2^n})^\circ)$$

is valid.

Proof. Based on theorem 44 we have to calculate the sum of degrees of all complex irreducible characters. Let $G := SD_{2^n}$, h an element of G, $o(h) = 2^{n-1}$ and a an involution in G such that $\langle h, a \rangle_\mathcal{G} = G$ and $h^a = h^{-1} \cdot h^{2^{n-2}}$ are valid. We define $z := h^{2^{n-2}}$. The conjugacy classes of G are the following ones (see exercises): $\{1\}$, $\{z\}$, a^G, $(ha)^G$ and for all $r \in \underline{2^{n-2} - 1}$, the class $\{h^{-r}, h^{-r}z\}$. Hence the class number $k(G)$ is exactly $2^{n-2} + 3$. h is generating a cyclic normal subgroup of index 2. Hence the degrees of all complex irreducible characters are 1 or 2 using a theorem of Ito. The quantity of all irreducible complex linear characters is exactly the index of the derived group. It is well-known (see exercises) that this index is 4. We conclude that there are 4 irreducible complex characters of degree 1 and $k(G) - 4$ irreducible complex characters of degree 2. Hence the dimension of all Cartan subalgebras of $(KG)^\circ$ is exactly $4 + (2^{n-2} + 3 - 4) \cdot 2 = 2 + 2^{n-1}$. Now we can finish the proof by specializing the corollaries 32 and 33 to a power of 2. ⋄

Corollary 35 *(direct product of two groups) Let K be a field and G, H finite groups. If KG and KH are semisimple, then $K(G \times H)$ is semisimple, too, and the identity*

$$cta((K(G \times H))^\circ) = cta((KG)^\circ) \cdot cta((KH)^\circ)$$

is valid.

Proof. $char(K)$ is zero or a prime number. By using the theorem of Maschke its straightforward to derive that $K(G \times H)$ is semisimple. We are focussing on the group algebra for the field $K = \mathbb{C}$ of G, H and $G \times H$. Let $\chi_1, \cdots, \chi_r$ and $\psi_1, \cdots, \psi_l$ be the complex irreducible characters of G and H with the degrees $n_1, \cdots, n_r$ and $s_1, \cdots, s_l$. Hence KG resp. KH is isomorphic to $K^{n_1 \times n_1}, \cdots, K^{n_r \times n_r}$ resp. $K^{s_1 \times s_1}, \cdots, K^{s_l \times s_l}$. It is easy to prove that $K(G \times H)$ is isomorphic to tensor product of KG and KH (see exercises). By fundamental properties of the tensor product we conclude that $K(G \times H)$ is isomorphic to a direct product of the algebras $K^{n_i \times n_i}$ and $K^{s_j \times s_j}$. For an arbitrary K-Algebra A and $n \in \mathbb{N}$ the identity $A^{n \times n} \cong K^{n \times n} \otimes A$ is valid (see exercises). Hence $K(G \times H)$ is isomorphic to a direct product of the algebras $(K^{n_i \times n_i})^{s_j \times s_j}$. In addition, we can prove (see exercises) that $(K^{n_i \times n_i})^{s_j \times s_j}$ is isomorphic to $K^{n_i s_j \times n_i s_j}$. In summarizing these results we conclude that $K(G \times H)$ is composed into a direct product of matrix-algebras over K of dimensions $n_i s_j$. These are the degrees of the complex irreducible characters of $G \times H$. Their sum is $\sum_{i,j} n_i s_j = \sum_i n_i \cdot \sum_j s_j$, and the proof is finished. ◇

With a standard induction-argument we derive the following corollary:

Corollary 36 *(direct products) Let $n \in \mathbb{N}$, K be a field and $G, G_1, \cdots, G_n$ be finite groups.*

(i) *If $KG_1, \cdots, KG_n$ are semisimple, then $K(G_1 \times \cdots \times G_n)$ is semisimple, too, and the identity $cta((K(G_1 \times \cdots \times G_n))^\circ) = \prod_{i=1}^{n} cta((KG_i)^\circ)$ is valid.*

(ii) *If KG is semisimple, then $K(G^n)$ is semisimple, too, and the identity $cta((KG^n)^\circ) = cta((KG)^\circ)^n$ is valid.*

By $Syl(G)$ we denote the set of Sylow subgroups of a finite group G. A finite nilpotent group is a direct product of its Sylow subgroups. By using corollary 36 we derive for finite nilpotent groups:

Corollary 37 *(nilpotent groups) Let K be a field and G a finite nilpotent group such that KG is semisimple. Then for every Sylow subgroup P of G the group algebra KP is semisimple, too, and the identity*

$$cta((KG)^\circ) = \prod_{P \in Syl(G)} cta((KP)^\circ)$$

is valid. ◇

Corollary 38 *Let G be a finite p-group and K a field such that KG is semisimple. Then p is a divisor of $cta((KG)^\circ)$.*

Proof. If G is abelian, then the corollary is a direct consequence of corollary 29. Let G be non-abelian. The number of the linear characters is a power of p because this value is exactly the order of G/G' and G' and G are not equal. The center of G is a normal subgroup which is not equal to G. By a theorem of Ito the degree of every irreducible complex character of G is a divisor of the order of $G/Z(G)$. Hence these degrees are powers of p. We conclude that the sum of degrees of all irreducible characters is a sum of powers of p, and we derive that p is a divisor of this value. We finish the proof by using theorem 44. ⋄

Corollary 39 *(Slattery-algorithm for p-groups)* In this corollary the algorithm of M.C. Slattery (see [61]) is presented and used to determine the degrees of the irreducible complex characters of p-groups recursively. For this we will use sequences $(a_n)_{n\in\mathbb{N}_0}$ of natural numbers including zero. An addition and multiplication with another sequence $(b_n)_{n\in\mathbb{N}_0}$ is defined component-by-component. By $shift((a_n)_{n\in\mathbb{N}_0})$ we denote the sequence $(b_n)_{n\in\mathbb{N}_0}$ with the conditions $b_0 = 0$ and $b_i = a_{i-1}$ for all $i \geq 1$.
The algorithm calculates the sequence $(c_n)_{n\in\mathbb{N}_0}$, and c_i is the number of complex irreducible characters of degree p^i for all $i \in \mathbb{N}_0$. In addition, we can calculate the sum of all these degrees, too, which is by using theorem 44 the dimension of all maximal tori of KG (= Cartan subalgebra of $(KG)^\circ$). Let p be a prime number and G a finite p-group. We define $(c_n(G))_{n\in\mathbb{N}_0}$ by:

(i) If the condition $\mid G \mid \leq p^2$ is valid, then we define $(c_n)_{n\in\mathbb{N}_0} := (\mid G \mid, 0, 0, ...)$.

(ii) Let $\mid G \mid \geq p^3$ and N be a central subgroup of G of order p. We define $(c_n(G))_{n\in\mathbb{N}_0}$ by $(c_n(G/N))_{n\in\mathbb{N}_0} + (p-1)(c_n(G, N))_{n\in\mathbb{N}_0}$.

The sequence $(c_n(G, N))_{n\in\mathbb{N}_0}$ is defined as follows:

(i) If $\mid G/N \mid \leq p$, then we define $(c_n(G, N))_{n\in\mathbb{N}_0} := (\mid G/N \mid, 0, 0, ...)$.

(ii) Let $\mid G/N \mid \geq p^2$, M be a normal subgroup of G containing N, $\mid M/N \mid = p$ and $C := C_G(M)$. At first let M be cyclic. If M is central, then we define $(c_n(G, N))_{n\in\mathbb{N}_0} := p(c_n(G, M))_{n\in\mathbb{N}_0}$. In the other case we set $(c_n(G, N))_{n\in\mathbb{N}_0} := (c_n(C, M))_{n\in\mathbb{N}_0}$. Now let M be non-cyclic. Choose an element $y \in M \setminus N$ of order p and $x \in N$ of order p. If M is central, then we define $(c_n(G, N))_{n\in\mathbb{N}_0} := \sum_{j=1}^{p}(c_n(G/\langle yx^j\rangle, M/\langle yx^j\rangle))_{n\in\mathbb{N}_0}$. In the other case we define $(c_n(G, N))_{n\in\mathbb{N}_0} =: shift(c_n(C/\langle y\rangle, M/\langle y\rangle))_{n\in\mathbb{N}_0}$. ⋄

Let G be a finite group. By $Irr(G)$ we denote the set of all complex irreducible characters of G.

Corollary 40 *(minimal non-abelian p-groups)* Let p be a prime number, G a minimal non-abelian p-group and K a field such that KG is semisimple $(char(K) \neq p)$. By using lemma 3.2.6.1 in [74] we conclude that the maximal subgroups are exactly the centralizers of non-central elements of G. G is nilpotent and hence these subgroups are normal subgroups of index p – and they are abelian which we will use later on. Therefore the length of the corresponding conjugacy class is p. In summarizing this we calculate:

$$|G| = |Z(G)| + (k(G) - |Z(G)|) \cdot p$$

and

$$k(G) = \frac{|G| - |Z(G)|}{p} + |Z(G)|.$$

By a theorem of Ito we conclude that every non-linear irreducible character is of degree p (because there are abelian normal subgroups of index p). In addition, using a theorem of Knoche (because every non-central conjugacy class has length p) we derive that the derived subgroup is of order p. Hence the following calculation is valid:

$$\begin{aligned} cta((KG)^\circ) &= \\ \sum_{\chi \in Irr(G)} deg(\chi) &= \\ |G/G'| \cdot 1 + (k(G) - |G/G'|) \cdot p &= \\ \frac{|G|}{p} + (\frac{|G| - |Z(G)|}{p} + |Z(G)| - |G/G'|) \cdot p &= \\ \frac{|G|}{p} + |G| - |Z(G)| + |Z(G)| \cdot p - |G| &= \\ \frac{|G|}{p} + |Z(G)| (p-1) & \quad . \end{aligned}$$

The minimal non-abelian p-groups are classified by three types (see e.g. 3.2.6.3 in [74]):

Type 1:
In this case $G = Q_8$ is valid and the sum is 6.

Type 2:
In this case G is of order p^{r+s} (for some $r, s \in \mathbb{N}$) and the center of G has

the order $max\{p^{r-1}, p^{s-1}\}$. The sum is:

$$\begin{aligned} cta((KG)^\circ) &= \\ \sum_{\chi \in Irr(G)} deg(\chi) &= \\ \frac{|G|}{p} + |Z(G)|(p-1) &= \\ p^{r+s-1}(p-1)max\{p^{r-1}, p^{s-1}\} &= \\ (p-1)p^{max\{2r-2+s, 2s-2+r\}}. \end{aligned}$$

Type 3:
In this case G is of order p^{s+r+1} (for some $r, s \in \mathbb{N}$) and the center of G has the order $max\{p, p^{r-1}, p^{s-1}\}$. The sum is:

$$\begin{aligned} cta((KG)^\circ) &= \\ \sum_{\chi \in Irr(G)} deg(\chi) &= \\ \frac{|G|}{p} + |Z(G)|(p-1) &= \\ p^{r+s}(p-1)max{p, p^{r-1}, p^{s-1}} &= \\ (p-1)p^{max\{2r-1+s, 2s-1+r, r+s+1\}}. \end{aligned}$$

The relevance of this sum is based on theorem 44: its the dimension of all maximal torus of KG (= Cartan subalgebra of $(KG)^\circ$).⋄

Example 6 *(S_4)* The symmetric group S_4 possesses two irreducible complex characters of degree 1, one of degree 2 and two of degree 3. By using theorem 44 the dimension of every Cartan subalgebra of $(KS_4)^\circ$ – K a field with $char(K) \neq 2, 3$ – is exactly $2 \cdot 1 + 1 \cdot 2 + 2 \cdot 3 = 10$: $cta((KS_4)^\circ) = 10$.⋄

Example 7 *(A_4)* The alternating group A_4 possesses 3 irreducible complex characters of degree 1 and one of degree 3. By using theorem 44 the dimension of every Cartan subalgebra of $(KS_4)^\circ$ – K a field with $char(K) \neq 2, 3$ – is exactly $3 \cdot 1 + 1 \cdot 3 = 6$: $cta((A_4)^\circ) = 6$.⋄

Both examples imply that the dimension of the maximal tori of $\mathbb{C}A_n$ and $\mathbb{C}S_n$ or not identical for an arbitrary n. An explicit formula for this dimension for the alternating groups is not known by the author as well as the difference of this dimension to the one of the symmetric groups. For the symmetric and generalized dihedral groups we will calculate this dimension by usage of the well-known Frobenius-Schur indicators. A group is called ambivalent if and only if every element is conjugated to its inverse.

Corollary 41 *(ambivalent groups)* If G is a finite ambivalent group and K a field such that KG is semisimple, then the identity
$$cta((KG)^\circ) = 1 + t(G)$$
is valid.

Proof. For ambivalent groups the well-known Frobenius-Schur indicator is exactly 1. Hence the 2nd root number function is the sum of all irreducible complex characters. The value of the 2nd root number function at 1 is $1+t(G)$. By using theorem 44 we finish the proof (for Frobenius-Schur indicators and nth root number functions see e.g. [27]).⋄

For the symmetric groups the so-called telephone numbers are related to the number of involutions (see e.g. [79]).

Corollary 42 *(symmetric groups)* Let $n \in \mathbb{N}$ and K be a field such that KS_n is semisimple. Then the following identity is valid:
$$cta((KS_n)^\circ) = \sum_{k=0}^{\lfloor \frac{n}{2} \rfloor} \frac{n!}{2^k(n-2k)!k!}.$$

Proof. We start the argumentation by proving that S_n is ambivalent. For every permutation the so-called cycle type is associated. For this, the permutation is presented as a product of disjoint cycle permutations. This decomposition is unique except the ordering of the factors. The length of all factors of this cycle decomposition are sorted ascending. The corresponding word is a partition of n – the cycle type. Two permutations are conjugate if and only if they possesses the same cycle type. In addition, the cycle type for a permutation and its inverse permutation is identical. Therefor S_n is ambivalent (see e.g. [6]).
By using corollary 41 and theorem 44 we have to calculate the value $1+t(S_n)$. This value is exactly the nth telephone number which is well-known as presented (see e.g. [79]).⋄

Dihedral groups are ambivalent groups, too (see exercises). In our context we study the generalized dihedral groups. Those groups are semidirect products of an abelian normal subgroup and a subgroup generated by an involution. The involution inverts all elements of the normal subgroup by conjugation. We symbolize these groups by $G = (A, \tau)$.

Let A be a finite abelian group. Then the set of all elements of order not greater than 2 form a elementary-2-abelian subgroup. This is obviously the greatest elementary-2-abelian subgroup symbolized by $\Omega_2(A)$. The determination of $\Omega_2(A)$ can be done based on a direct composition of A in cyclic groups. This is to be done by the reader as an exercise.

Corollary 43 *(generalized dihedral group)* Let $G = (A, \tau)$ be a generalized dihedral group and K a field such that KG is semisimple. Then the following identity is valid:

$$cta((KG)^\circ) = \mid \Omega_2(A) \mid + \mid A \mid.$$

Proof. We begin the argumentation by proving the ambivalence of G. Let t be an involution acting invertible on A by conjugation. The index of A in G is 2 and hence G is the disjoint union of A and tA. By definition each element of A is conjugated by t to its inverse. Let a in A. Then the identity $tata = t^{-1}ata = a^{-1}a = 1$ is valid. Therefor all elements of tA are involutions which are identical (and hence conjugated) to its inverse.
By using corollary 23 we have to count all elements of order ≤ 2. All elements of tA are involutions and their quantity is the same as the order of A. Within A we have done this calculation just before this corollary.⋄

For the following corollary modern results for linear groups are used to calculate the dimension of maximal tori in group algebras. The result for the symmetric groups was based on the theory of Frobenius-Schur indicators. The next results use the theory of twisted Frobenius-Schur indicators and symmetric functions.

Let K be a field and $n \in \mathbb{N}$. The general linear group of degree n over K – denoted by $GL(n, K)$ – is the set of all invertible $n \times n$-matrices over K. If K is finite – and $\mid K \mid$ a prime power q –, then we symbolize this group by $GL(n, q)$. Its order is $\prod_{i=0}^{n-1}(q^n - q^i)$. By using results of Rhodes and Macdonald we are able to derive the following corollary:

Corollary 44 *(general linear group)* Let $n \in \mathbb{N}$, K a finite field of order $q := p^k$, p a prime number, $k \in \mathbb{N}$ and $G := GL(n, q)$. Let F be a field such that FG is semisimple. The dimension of a maximal torus of FG is exactly the number of invertible symmetric matrices in G. This value is to be calculated by the formula:

$$cta((KG)^\circ) = (p^k)^{\binom{n}{2}} \prod_{i=0}^{\lfloor \frac{n}{2} \rfloor} (1 - (\tfrac{1}{p^k})^{2i-1}).$$

In addition, the product $\prod_{i=0}^{\lfloor \frac{n}{2} \rfloor}(1 - (\tfrac{1}{p^k})^{2i-1})$ is the probability that a symmetric $n \times n$-matrix is invertible.

Proof. I.G. Macdonald proved by using symmetric functions (see [45], chapter IV.6, Ex. 5) that the sum of degrees of all irreducible complex characters is exactly the number of invertible symmetric matrices. This value is calculated in [51] as presented. By using theorem 44 we finish the proof. The add-on is valid by using a result in [51].⋄

Let $n \in \mathbb{N}$, q an uneven prime power and $K := GF(q)$. The symplectic group of degree $2n$ over K is symbolized by $SP(2n,q)$. An exact definition is presented within [86]. By using so-called twisted Frobenius-Schur indicators Vinroot calculated the sum of degrees of all irreducible complex characters of G:

Corollary 45 *(symplectic group) Let $n \in \mathbb{N}$, K a finite field of order $q := p^k$, p an uneven prime number, $k \in \mathbb{N}$ and $G := SP(2n,q)$. Let F be a field such that FG is semisimple. The dimension of a maximal torus of FG is exactly the index of $GL(n,q)$ in G. This value is to be calculated by the formula:*

$$cta((KG)^\circ) = q^{\frac{1}{2}n(n+1)} \cdot \prod_{i=0}^{n}(q^i+1).$$

Proof. The sum of all irreducible complex G-characters is calculated by Vinroot in [70] (theorem 1.3, corollary 6.1) as presented. By using theorem 44 we finish the proof.⋄

Vinroot proved in [70] that this value is exactly the number of skew-symmetric matrices in $GSP(2n,q)$ – the general similitude group. In [70] he calculated the sum of the degrees of all irreducible complex characters for $GSP(2n,q)$, too. We derive the following result using theorem 44 ($U(n,q)$ is the unitary group):

Corollary 46 *(general similitude group) Let $n \in \mathbb{N}$, K a finite field of order $q := p^k$, p an uneven prime number, $k \in \mathbb{N}$ and $G := GSP(2n,q)$. Let F be another field such that FG is semisimple. The dimension of a maximal torus of FG is exactly the number of symmetric matrices of G. This value is to be calculated by the formula:*

$$cta((KG)^\circ) = \frac{|G|}{2|GL(n,q)|} + \frac{|G|}{2|U(n,q^2)|}.\diamond$$

Vinroot proved in [70] the identity

$$(q^{\frac{1}{2}n(n+1)} \cdot \prod_{i=0}^{n}(q^i+1)) \cdot (q-1)\cdot \mid GL(n,q) \mid = \mid GSP(2n,q) \mid.$$

Within his articles [71] and [72] he studied orthogonal groups, too. The reader might study these articles as an exercise and derive a summation formula for these groups, too.

Example 8 *(ax + b-group)* Let K be a finite field with $q \geq 3$ elements, p a prime number and $n \in \mathbb{N}$ with $q = p^n$. For all $0 \neq a, b \in K$ we define the function $\tau_{a,b} : x \mapsto ax + b$, and G to be the set of all these functions: the $ax+b$-group. The set of functions $\tau_{1,b}$ is a normal subgroup and coincide with the derived subgroup. The derived subgroup is isomorphic to the additive group of the field K. In addition, the set of functions $\tau_{a,0}$ is a complement of the derived group. The complement is isomorphic to the multiplicative group of the field K. The $ax + b$-group has exactly q conjugacy classes. Therefor the group possesses exactly $q - 1$ linear complex characters and one irreducible complex character of degree at least 2. The sum of squares of the degrees of all irreducible complex characters is $q(q - 1)$ – the order of the group. Hence we have exactly one non-linear irreducible complex character of degree $q - 1$. The sum of all degrees of the irreducible complex characters is exactly $2q - 2$. This value (see theorem 44) is the dimension of the maximal tori (= Cartan subalgebras) in every semisimple group algebra related to the $ax + b$-group: $cta((KG)^\circ) = 2q - 2$.◇

The $ax + b$-group is a Frobenius group. They are defined in the following way: there exists a non-trivial subgroup H of G such that for all $g \in G \setminus H$ the intersection of H and H^g is the trivial subgroup. The subgroup H is called the Frobenius complement. The Frobenius kernel C – which leads to a semidirect product of G together with the complement (which is the theorem of Frobenius) – is defined by $C := (G \setminus \bigcup_{g \in G} H^g) \cup \{1\}$.

The character theory of Frobenius groups for semisimple group algebras is well-known. Therefor we will extend our example of the $ax + b$-group to arbitrary Frobenius groups in the next corollary. The previous example is to be re-calculated by the reader in the exercises as an easy consequence of the next corollary. The sum will be reduced to the correspondent one of the kernel and complement:

Corollary 47 *(Frobenius groups)* Let G be a Frobenius group with kernel C and complement H and let K be a field such that KG is semisimple. Then the identity

$$cta((KG)^\circ) = cta((KH)^\circ) + cta((KC)^\circ) - 1$$

is valid. In addition, if C and H are abelian, then the formula is reduced to:

$$cta((KG)^\circ) = |C| + |H| - 1.$$

Proof. We use theorem 16.13 related to the character theory of Frobenius groups in [25].
There are $k(H)$ irreducible complex characters μ_i which are the extensions of all irreducible complex characters of H for G defined by $\mu_i(ch) = \mu_i(h)$ for all $h \in H$ and $c \in C$. The degrees of these extensions are the same as

for the original H-characters. In particular, their sum is the same as for the original ones.

All other irreducible G-characters are to be constructed in the following way: for all $k(C) - 1$ non-trivial irreducible C-characters ψ the induced character ψ^G of G is irreducible, too. The degree of the induced character is the product of the degree of the original character with the index of the correspondent subgroup. In this case this index is the one of C which is identical to the order of H. Two induced characters are identical if and only if they are conjugated under H. For one ψ there are exactly $\mid H \mid$ conjugates which have the same degree. The sum of degrees of these $\frac{k(C)-1}{|H|}$ irreducible complex characters is therefor to be calculated by the following formula:

$$\sum_{i=1}^{\frac{k(C)-1}{|H|}} grad((\psi_i)^G)) =$$
$$\sum_{i=1}^{\frac{k(C)-1}{|H|}} grad(\psi_i)) \cdot \mid G/C \mid =$$
$$\sum_{i=1}^{\frac{k(C)-1}{|H|}} grad(\psi_i) \cdot \mid H \mid =$$
$$\sum_{i=1}^{\frac{k(C)-1}{|H|}} \sum_{h \in H} grad((\psi_i)^h) =$$
$$\sum_{i=1}^{k(C)-1} grad(\psi_i) =$$
$$cta((KC)^\circ) - 1 \quad .$$

This is the sum of all degrees of the irreducible complex non-trivial C-characters. The degree of the trivial one is 1. (The irreducible characters are labeled such that the highest index corresponds to the trivial C-character.) By using theorem 44 and corollary 29 we finish the proof.⋄

We remark the following results: Thompson (see e.g. [25]) proved that the kernel of a Frobenius group is nilpotent. Therefor the calculation for the kernel can be reduced to direct products of p-groups and hence by corollary 36 to p-groups directly. For them we can apply the Slattery-algorithm (see corollary 39). The structure of the complement of Frobenius groups is limited: in case of an uneven order the complement is meta-cyclic. These groups are handled in corollary 49. In fact, the kernel is abelian in this case, too. A limitation for the other case is not known by the author.

Frobenius groups are special semidirect products. For another special case of semidirect products – with abelian normal subgroup – the construction of the irreducible complex characters is well-known. The reader can study the correspondent part in the textbook of Serre ([57], Part II, Section 8.2.). The author thanks Alex Bartel for his support (by mail and in the forum Mathstackexchange [87]).

Corollary 48 *(Semidirect products with abelian normal subgroup)* In this part we analyze semidirect products $G = AH$ with an abelian normal subgroup A and a complement H. H is acting on the set of irreducible complex characters (likewise presented for the Frobenius groups, too!) by conjugation. Let χ be an irreducible complex character of A (which is a linear one by the commutativity of A). We denote the stabilizer of χ in H by $Stab_H(\chi) := \{h \mid h \in H, \chi^h = \chi\}$. It is a subgroup of H. We focus on the semidirect product $S_\chi = A \rtimes Stab_H(\chi)$. We extend χ on S_χ by $\hat{\chi}(as) := \chi(a)$ for all $a \in A$ and $s \in Stab_H(\chi)$. Let ρ be an irreducible complex character of $Stab_H(\chi)$. Then $(\hat{\chi} \otimes \rho)^G$ is an irreducible complex character of G. In addition, all irreducible complex characters are determinable by this extension process. Their degree is $\mid H/Stab_H(\chi) \mid \cdot grad(\rho)$. Not all these extensions are in-equivalent: the Mackey-formula is usable to decide whether two extensions $(\hat{\chi_1} \otimes \rho_1)^G$ and $(\hat{\chi_2} \otimes \rho_2)^G$ are equivalent: χ_1 and χ_2 must be conjugated under H. Therefor the correspondent stabilizers $Stab_H(\chi_i)$ and the subgroups S_{χ_i} are conjugated under H, too. This argumentation is used to calculate the sum of degrees of irreducible complex characters. We determine this sum in general and for one example. There are some more examples related to this topic for the reader provided in the exercises.

Let $B_1, ..., B_l$ be the orbits of H by conjugation on $Irr(A)$ and let $\chi_i \in B_i$ for all $i \in \underline{l}$ be a system of distinct representatives of these orbits. l is the number of distinct orbits of $Irr(A)$ under H. The irreducible complex characters $\chi_i \in B_i$ for $i \in \underline{l}$ are pairwise not conjugated. Let $i \in \underline{l}$. If $\rho_{i_1}, ..., \rho_{i_{r_i}}$ are the irreducible complex characters of $Stab_H(\chi_i)$, then $(\hat{\chi_i} \otimes \rho_{i_s})^G$ for $s \in \underline{r_i}$ are the pairwise distinct irreducible complex characters of G. For every $j \in \underline{l}$ we create by this process always new irreducible complex characters of G.

Their sum of degrees can be calculated by the formula:

$$
\begin{aligned}
cta((KG)^\circ) &= \\
\sum_{\chi \in Irr(G)} deg(\chi) &= \\
\sum_{i=1}^{l}\sum_{s=1}^{r_i} deg(\hat{\chi}_i \otimes \rho_{i_s})^G &= \\
\sum_{i=1}^{l}\sum_{s=1}^{r_i} |H/Stab_H(\chi_i)| \cdot deg(\rho_{i_s}) &= \\
\sum_{i=1}^{l} |H/Stab_H(\chi_i)| \sum_{s=1}^{r_i} deg(\rho_{i_s}) & \quad .
\end{aligned}
$$

Hence we have reduced the calculation of the calculation of the correspondent sum of degrees of the stabilizers in H. If H is commutative, then the sum is to be determined by the formula:

$$
\begin{aligned}
cta((KG)^\circ) &= \\
\sum_{\chi \in Irr(G)} deg(\chi) &= \\
\sum_{i=1}^{l} |H/Stab_H(\chi_i)| \sum_{s=1}^{r_i} deg(\rho_{i_s}) &= \\
\sum_{i=1}^{l} |H/Stab_H(\chi_i)| \sum_{s=1}^{r_i} 1 &= \\
\sum_{i=1}^{l} |H/Stab_H(\chi_i)| \cdot |Stab_H(\chi_i)| &= \\
\sum_{i=1}^{l} |H| &= \\
l \cdot |H| & \quad .
\end{aligned}
$$

By using theorem 44 this value is the dimension of a maximal torus (= Cartan subalgebra) of KG for every field K such that KG is semisimple.

As an example we study the semidirect product G of Z_p with Z_n such that $n \mid p-1$ is valid and Z_n acts faithful on Z_p. In this faithful case every non-trivial irreducible complex character acts with a trivial stabilizer and the one for the trivial character is exactly Z_n. Z_p possesses p linear and $p-1$ non-trivial characters. Every character among these $p-1$ is related to the orbit size n under Z_n. Hence we have $1 + \frac{p-1}{n}$ orbits under Z_n. The sum of degrees can be calculated by:

$$cta((KG)^\circ) = \sum_{\chi \in Irr(G)} deg(\chi) = l \cdot |H| = (1 + \tfrac{p-1}{n}) \cdot n = n + p - 1. \diamond$$

Corollary 49 *(meta-cyclic groups)* In the argumentation within this corollary we use the main results within the articles [9] and [3] about meta-cyclic groups. Let G be a finite meta-cyclic group (a finite group which possesses a cyclic normal subgroup such that its factor group is cyclic, too). There are prominent examples among this class of groups: groups of a square-free order and Z-groups which are finite groups possessing only cyclic Sylow subgroups.

A meta-cyclic group G can be presented in the following way:
There exist $n, m, t, r \in \mathbb{N}$ and $a, b \in G$ such that G is generated by a, b and is of order nt. In addition, the following conditions are valid:
$o(a) = n$, $o(b) = m$, $a^k = b^t$, $a^b = a^r$, $t \mid m$, $r^t \cong 1 \bmod n$, $kr \cong k \bmod n$ and $gcd(r, n) = 1$.

The representation theory is known as well:
Let U_n be the set of complex nth-roots of unity, σ the powering with r on U_n, $t(\alpha)$ the length of the orbit of $\alpha \in U_n$ with respect to $\langle \sigma \rangle$ and $o(\alpha)$ the order of $\alpha \in U_n$. By using [3], pages 15ff., all irreducible complex characters are created with respect to $\alpha \in U_n$ and some special Θ – noted by $T_{\alpha, \Theta}$. Their degree is $t(\alpha)$. $T_{\alpha, \Theta}$ is $t(\alpha)$-times equivalent if α is varying. For every α there are $\frac{t}{t(\alpha)}$ distinct Θ.
Therefor we can derive for the sum of degrees of all complex irreducible characters by the formula:

$$cta((KG)^\circ) = \sum_{\chi \in Irr(G)} deg(\chi) = \sum_{\alpha \in U_n} t(\alpha) \tfrac{t}{t(\alpha)^2} = (\sum_{\alpha \in U_n} \tfrac{1}{t(\alpha)}) t.$$

In [3] this sum of $t(\alpha)$ is re-calculated which is usable for our needs, too. Let ϕ be Euler's totient function and $s(d)$ the order of $r + Zd$ in $E(Z/Zd)$ for all divisors d of n. We conclude:

$$cta(((KG))^\circ) = \sum_{\chi \in Irr(G)} deg(\chi) = (\sum_{\alpha \in U_n} \tfrac{1}{t(\alpha)}) t = (\sum_{d \mid n} \tfrac{1}{\phi(d) s(d)}) t.$$

This value is (with respect to theorem 44) the dimension of a maximal torus (= Cartan subalgebra) of KG for every field K such that KG is semisimple. $\diamond$

6.4 The modular case

Based on theorem 44 we can derive another theorem in the case of modular group algebra calculating the dimension of maximal tori. The next lemma is used for the proof of this result and will play an important role in chapter 7, too:

Lemma 11 *Let K be a field, A an associative finite-dimensional unitary K-algebra and N a nilpotent ideal of A. If T is a maximal torus of A, then $(T+N)/N$ is a maximal torus of A/N.*

Proof. A torus possesses no non-zero nilpotent elements. By using fundamental isomorphism theorems of algebras T and $(T+N)/N$ are isomorphic. If D be a torus of A/N containing $(T+N)/N$, then there exists a subalgebra S (containing N) of A such that $D = S/N$ is valid. The algebra S is solvable, its nilradical is exactly N (because N is nilpotent and S/N is separable), and the theorem of Wedderburn-Malcev implies the existence of a separable subalgebra X in S isomorphic to D. X and D are isomorphic, and hence X is a torus, too. T is separable and contained in S and thus – using again the theorem of Wedderburn-Malcev – can be conjugated into X. T is a maximal torus, and therefor X and T have the same dimension. We conclude that the identity $D = (T+N)/N$ is valid. ⋄

Theorem 45 *(second summation formula by Salvatore Siciliano) Let G be a finite group, K a field of characteristic p dividing the order of G, P a normal p-Sylow subgroup of G and $\psi_1, \psi_2, \ldots, \psi_t$ the irreducible complex characters of G which are trivial on P. The dimension of all maximal tori of KG is determined by the summation formula $\sum_{i=1}^{t} \deg(\psi_i)$. (This is also the sum of all irreducible complex characters of G/P.)*

Proof. Let T be a maximal torus of KG. By using lemma 11 the algebra $(T+rad(KG))/rad(KG)$ is a maximal torus of $KG/rad(KG)$. The kernel of the linearization of the natural epimorphism of G onto G/P is exactly (see e.g. [49]) $Aug(KP)KG = KGAug(KP)$. By using a theorem of Wallace we derive that $Aug(KP)$ is nilpotent and hence the kernel is nilpotent, too. The factor algebra of KG with respect to this kernel is isomorphic to $K(G/N)$. A theorem of Maschke ensures that $K(G/P)$ is semisimple (because p is not dividing the order of H). We conclude that the kernel is exactly the nilradical. We apply theorem 44 to the semisimple algebra $K(G/P)$. The nilradical of KG contains no non-zero separable element. Hence $T \cap rad(KG)$ is the null-space and the algebras T and $(T+rad(KG))/rad(KG)$ are isomorphic. The dimension of T is the sum of the degrees of the complex irreducible characters of $K(G/N)$. Using [27], chapter 15, these degrees are exactly the same as the ones of ψ_i. ⋄

Example 9 *(Siciliano)* When the characteristic of the ground field divides the order of a given finite group, it is not always true that the dimension of a Cartan subalgebra is the sum of the degrees of the irreducible characters (defined over $\mathbb{C}$), not even under the hypotheses of theorem 45. For example, consider the group G generated by a, b, c, d subject to the relations

$$a^2 = b^3 = c^3 = d^3 = 1;$$
$$ca = ac^2; \qquad cb = bcd;$$
$$da = ad^2; \qquad ab = ba;$$
$$bd = db; \qquad cd = dc.$$

This group has order 54. Let $F = \mathbb{F}_3$. Then $FG \cong T \oplus rad(FG)$ with T a maximal torus of dimension 2, and $rad(FG)$ of dimension 52. In this case $H = C_{FG}(T)$ has dimension 30. Let N be the subgroup of G generated by b and c. Then N is a normal 3-Sylow subgroup of G but

$$\sum_{\chi \in Irr(G)} \deg \chi = 18.$$

Note also that H is not a maximal torus of FG. (The computations for this example have been done by Salvatore Siciliano using the computer algebra system GAP [82]).⋄

Corollary 50 *Let p be a prime number, K a field of characteristic p, G a finite group and P a p-group such that p is not dividing the order of G. The dimension of the maximal tori of $(K(G \times P))°$ is the same as the dimension of the Cartan subalgebras (and maximal tori) of $(KG)°$.*

Proof. The proof is a direct consequence of the theorems 44 and 45 because $(G \times P)/P$ and G are isomorphic. ⋄

Corollary 51 *Let p be an uneven prime number, K a field of characteristic p and $n \in \mathbb{N}$. Every maximal torus of $(KD_{2p^n})°$ has dimension 2.*

Proof. The proof is based on theorems 44 and 45 as well as on corollary 29: the group D_{2p^n} possesses a normal p-Sylow subgroup of order p^n and index 2. Its factor group is abelian of order 2. ⋄

Chapter 6

Closure of Groups

topic	formula
extra special of order p^{2n+1}	$p^n(p^n+p-1)$
D_{2n}, n even	$n+3$
n uneven	$n+2$
Q_{4n}	$2n+2$
SD_{2^n}	$2^{n-1}+2$
S_4	10
A_4	6
6 abelian	$\|G\|$
$\|G\| = p^2$	$p+q-1$
(non abelian)	
bounds relevant	$1 + t(G)$
p-groups	"$\text{Irr}(G)$" "class"
Cameron-Rockert	by theorem S/low
unipotent Y	subgroup
$\text{GL}_n \times S$ over $G_t(q)$	$2q-2$

Bounds

topic	formula	bound
universal bound, abelian p-group	$\frac{\|G\|}{p} + (p-1) \cdot \lceil 2(G) \rceil$	$\leq \|G\|$
generalization of direct products of n copies of S_n	$\|A\|+\|O_2(A)\|$	$\geq \sqrt{\|G\|}$
S_n		$\geq \|A\|$
$G = C_n \rtimes \text{Frob}_{1,n}$	by theorem number	$\geq 1 + t(G)$
$G = A \ltimes H$	sum of degrees	$\geq 2\sqrt{\|G\|-1}$
$\|G\|$	$\|A\|$(sum of orb. of C acts $H-1$)	$\geq \|G\| < 1$
$G = P \ltimes Q$	with $\|H\|$ a conj.	$=\sqrt{\|G\|}+1-\frac{\|G\|}{|P|}$
	$t(\sum \frac{1}{|C_{G}(x)|^2 / |C|})$	$= \sqrt{(\|G\|-1)/\|G\|(\|G\|)(\|G\|)}$
p-group	unipotent	$\leq \sqrt{(\|G\|-1)\|G\|+|\text{Z}(G)|}$
	classes, Z.a.d	$-\frac{\|G\|}{\omega\sqrt{|G|}}$
		$p^4 | |G| $
		$\geq p$, $w\neq \pi$
		$r = \lfloor \sqrt{8q+1} - 1 \rfloor$

6.5 Open-ended questions and exercises

Open-ended questions 10 *(i) How can we calculate the sum of the degrees of the complex irreducible characters for group classes like the alternating groups and the sporadic simple groups?*

(ii) How can we construct all maximal tori and Cartan subalgebras for $(KG)°$ in the modular and semisimple case?

(iii) Is every natural number a dimension of a maximal torus or a Cartan subalgebra in a specific semisimple/modular group algebras for a non-abelian group?

(iv) Are wreath products of ambivalent groups ambivalent, too? What is the number of involutions for these ambivalent wreath products?

(v) What is the number of involutions for the wreath product of a finite ambivalent group with a symmetric group? (A. Kerber proved that this wreath product is ambivalent.)

(vi) Determine the sum of the degrees of the complex irreducible characters for central products with respect to their factors!

(vii) Determine the sum of the degrees of the complex irreducible characters for products of united factor groups with respect to their factors!

(viii) The representation theory of wreath products is analyzed by Adalbert Kerber in the Canadian J. Math. 20(1968), 665-672. Is it possible to calculate the sum of all complex irreducible characters based on his analysis? Is there a special formula for this sum if we restrict the problem to regular wreath products of cyclic groups?

(ix) How can we calculate the sum of all complex irreducible characters for Frobenius groups with a complement of even order?

Excercise 254 Study the articles [43] and [69] and construct a maximal torus in the complex group algebra of the symmetric group by using the so-called Murphy elements of the group algebra of the symmetric group.

Excercise 255 Is it possible that a generalized dihedral group is a Frobenius group? If the answer is 'yes', then provide an example!

Excercise 256 Let G be a group of order $3 \cdot 5 \cdot 7$. Is G meta-cyclic? If the answer is 'yes', then use the results of this chapter for meta-cyclic groups on this specific example.

Excercise 257 Use the algorithm of M.C. Slattery 39 on an extra-special p-group of order p^3. Is it possible to extend the result for an arbitrary extra-special p-group?

Excercise 258 Let K be a field and G a non-abelian p-group such that KG is semisimple. G possesses a cyclic maximal subgroup. What kind of results can you derive by using theorem 44 for these groups? (Tip: these groups are classified, see e.g. [63]).

Excercise 259 The group G possesses a center (as large as possible) of index p^2. What are the consequences of theorem 44 for these groups? (Tip: At first prove that all non-central conjugacy classes are of length p. Now use a result of Knoche to determine the derived subgroup of G. Prove further that G possesses an abelian maximal subgroup which is a normal subgroup, too. Determine the class number in terms of $\mid Z(G) \mid$ and $\mid G \mid$. Use a result of Ito to bound the degrees of the complex irreducible characters and hence their sum, too. A similar argumentation was used in this chapter for minimal non-abelian p-groups.)

Excercise 260 Transfer the argumentation of exercise 259 to the following p-groups: the derived subgroup is of order p and the group possesses a maximal abelian subgroup!

Excercise 261 By using corollary 48 determine the sum of all degrees of the irreducible characters in their complex group algebras and derive meaningful results with theorem 44:

(i) $Z_p \wr Z_p$, p a prime number

(ii) $Z_2 \wr Z_p$, p a prime number

(iii) $Z_2 \wr Z_n$, n a natural number (finite lamplighter group)

(iv) $Z_n \wr Z_n$, n a natural number

(v) $Z_n \wr Z_m$, n, m natural numbers

(vi) $A \wr Z_p$, p a prime number and A a finite abelian group.

(Tip: If the general case for p, n, m is to difficult to solve, then try to study the problem with small values. In addition, it might be helpful to deal with direct products of cyclic groups to handle A.)

Excercise 262 Vinroot has studied in his articles [71] and [72] the orthogonal groups and determine a formula for the sum of all degrees of the irreducible complex characters and hence for the dimension of maximal tori. Read and study his articles and derive a result with theorem 44 (as done for the other linear groups in this chapter).

Excercise 263 For prime numbers $p, q \leq 31$ calculate – by using corollaries 30 and 31 – the sum of all degrees of the irreducible characters in their complex group algebras for groups of order pq, p^3 and q^3. What is the structural meaning of this value? Which fields are suitable for their group algebra being semisimple? Summarize this exercise in a table. Can you answer the question for arbitrary prime numbers p, q, too?

Excercise 264 What are the consequences of theorem 44 for the regular wreath product of Z_2 with S_4? Is this group ambivalent?

Excercise 265 Calculate the sum of all degrees of the irreducible characters of $\mathbb{R}G$ for the following cases of the group G (p a prime number, $n \in \mathbb{N}$):

(i) $GL(2,2)$, $SP(4,2)$, $GSP(4,2)$

(ii) $GL(2,p)$, $SP(4,p)$, $GSP(4,p)$

(iii) $GL(p,p)$, $SP(2p,p)$, $GSP(2p,p)$

(iv) $GL(n,2)$, $SP(2n,2)$, $GSP(2n,2)$

(v) $GL(2,p^p)$, $SP(4,p^p)$, $GSP(4,p^p)$

(vi) $GL(3,5)$, $SP(6,5)$, $GSP(6,5)$.

Excercise 266 Let p be a prime number. Use theorem 44 to derive meaningful results for the regular wreath product of Z_p with Z_p an. (Tip: The derived subgroup is of index p^2, the group possesses an abelian normal subgroup of index p and the group is of order p^{p+1}.)

Excercise 267 For the following groups derive from corollary 48 with a suitable direct composition of the group meaningful results: groups of order pq^2 ($p, q \in \mathbb{P}$), S_3, S_4, A_4 and D_{2n} ($n \in \mathbb{N}$).

For solving this exercise do a research in the literature to determine the structure and irreducible complex characters of groups of order pq^2 for prime numbers p, q.

Excercise 268 True or false: Finite meta-cyclic groups are nilpotent, supersolvable and meta-abelian.

Excercise 269 Under what terms are dihedral, semi-dihedral and quaternion groups meta-cyclic?

Excercise 270 Under what terms is a finite abelian group meta-cyclic?

Excercise 271 Prove all statements within example 8!

Excercise 272 Prove example 8 again but now by using theorem 47. In addition, prove that the $ax + b$-group is a Frobenius group!

Excercise 273 True or false: A_n and S_n ($n \leq 4$) are Frobenius groups. Apply theorem 47 on them! What is the answer for $n \geq 5$?

Excercise 274 In [25], examples 8.6, pages 498-499 there are three examples of Frobenius groups. Apply corollary 47 to these groups!

Excercise 275 Do a research in the literature and define the term M-group. What results are derivable by using theorem 44 for M-groups? (Tip: What is the dimension of an induced character?)

Excercise 276 The set of all invertible 2×2 upper triangular matrices with determinant 1 with respect to a finite field with at least three elements is a Frobenius group. The subgroup of diagonal matrices is a Frobenius complement and the subgroup of strict upper triangular matrices with diagonal entries 1 is a Frobenius kernel. Prove this statements and apply corollary 47 to this group!

Excercise 277 Compare the dimension of a maximal torus of $\mathbb{C}S_n$ and $\mathbb{C}A_n$ for $n = 1, 2, 3, 4, 5$. Is the value for S_n still not less than the one for A_n? Is there a result explaining this phenomenon for a group and their subgroups?

Excercise 278 Let P be a group of order p, p^2 or p^3 for an arbitrary prime number p. Calculate the dimension of the Cartan subalgebra in their complex group algebra.

Excercise 279 Let P be a group of order p^4 for an arbitrary prime number p. By using the article [88] calculate the dimension of the Cartan subalgebra in their complex group algebra. Is this dimension still not less than the dimension for a group of order p, p^2 or p^3?

Excercise 280 Let p be a prime number and P, H p-groups satisfying $|P| \geq |H|$. True or false: The dimension of a maximal torus of $\mathbb{C}P$ is not less than the one for $\mathbb{C}H$. What is the answer of this question if H is a subgroup of P? Is there a relation to induction or restriction of characters?

Excercise 281 Use the article [89] to study the dimension of a maximal torus for a complex group algebras related to central products. Is there a way to bound this dimension by the corresponding dimensions of the factors? (Tip: Use the article to determine the degrees of the irreducible characters!)

Excercise 282 For the symmetric group of degree $n \leq 20$ calculate the dimension of a maximal torus in the rational group algebra. Will the dimension change for the real or complex group algebra? What is the answer of this question for a finite field such that the group algebra is still semisimple? Describe or determine these finite fields for $n \leq 20$!

Excercise 283 *True or false: Subgroups, normal subgroups, factor groups and direct products of ambivalent groups are ambivalent, too.*

Excercise 284 *Prove the recursion formula $T(n) = T(n-1) + (n-1) \cdot T(n-2)$ for the telephone numbers $T(n)$ and derive the explicit formula for them presented in this chapter!*

Excercise 285 *Prove that the set of involution united with the unit element is the greatest elementary-2-abelian subgroup of a finite abelian group. Is this result true for an arbitrary finite group?*

Excercise 286 *Using the results of this chapter for a generalized dihedral group derive the corresponding ones for D_{2n}. What is the greatest elementary-2-abelian subgroup of a finite cyclic group?*

Excercise 287 *Let $G = (A, \tau)$ be a generalized dihedral group. What is the consequence for its structure if the normal subgroup A is elementary-2-abelian?*

Excercise 288 *Under what terms is a finite abelian group ambivalent?*

Excercise 289 *Every finite nilpotent ambivalent group is a 2-group. Is every 2-group ambivalent?*

Excercise 290 *Every finite solvable ambivalent group is a 2-group. A finite ambivalent group is solvable if and only if it is nilpotent.*

Excercise 291 *Under what terms are the following groups ambivalent: D_{2n}, SD_{2^n}, Q_{4n}? Under these terms calculate the dimension of the Cartan subalgebra in their complex group algebra! Is it possible to determine this dimension in the non-ambivalent case, too?*

Excercise 292 *Develop a procedure to calculate the greatest elementary-2-abelian subgroup of products of abelian groups!*

Excercise 293 *For a generalized dihedral group $G = (A, \tau)$ determine the sum of degrees of all irreducible complex representations for the following cases of A: $Z_2 \times Z_3$, $Z_5 \times Z_3 \times Z_2$, $Z_6 \times Z_8$. Determine a calculation for an arbitrary abelian group A represented as a product of cyclic groups! Emphasize the case of one and two factors!*

Excercise 294 *Let P be a p-group and A an abelian p-group. How can we calculate the sum of degrees of all irreducible complex representations of $P \times A$? Is the case of an extra-special p-group P to be handled differently?*

Excercise 295 Look-up the character tables in [27] for the following groups and determine their sum of degrees of all irreducible complex representations: $S_4, A_4, A_5, SL(2,3), SL(2,4), GL(2,3), A_6, SL(2,8), M_{11}$ and $PSL(2,11)$. Alternatively use suitable results of this chapter to solve this exercise!

Excercise 296 Take the Atlas of the finite simple sporadic groups. Calculate the sum of degrees of all irreducible complex representations. What is the result for the monster group?

Excercise 297 By a research in the literature (e.g. [83]) determine the character table of $SL(2,q)$ and $PSL(2,q)$ (q a prime power). Calculate the sum of all degrees of the irreducible characters? In what way is the value $2.5q + 0.5q^2 + 4$ important for this sum? What is the meaning of this sum for the Cartan subalgebras?

Excercise 298 Analyze the same question as in exercise 297 for the group $GL(2,q)$. In addition, use a theorem presented in this chapter for $GL(n,q)$ to solve this problem, too. In what way is the value $q^2(q-1)$ important for this problem?

Excercise 299 Let p be a prime number and P a p-group which possesses an abelian maximal subgroup A. Then the sum of all degrees of complex irreducible characters can be calculated by the formula $\mid A \mid - \mid G \mid + p \cdot k(G)$. (Tip: theorem of Ito, maximal subgroups of nilpotent groups). Specify prominent examples of this situation!

Excercise 300 Let (K, F) be a field extension and G a finite group. Prove that the F-Algebra $KG \otimes_K F$ is isomorphic to FG.

Excercise 301 The Kleinian group V_4 possesses no faithful complex irreducible representation. Specify a faithful V_4-representation of degree 2! Calculate the sum of degrees of all irreducible complex representations of V_4!

Excercise 302 Research the literature with respect to the following topic: Which finite groups possess a faithful irreducible complex representation? (Tip: Rudolf Kochendörffer and exercise 304).

Excercise 303 Using corollary 28 calculate the bounds for an arbitrary prime number p and $n \leq 30$! In addition proof for a prime number p: if p^{10} divides the order of a finite group, then the dimension of a maximal torus in the correspondent group algebra is at least p^4.

184

Excercise 304 *Perform a research on the Japanese mathematician Kenjiro Shoda[1] with respect to faithful meta-abelian groups. What is his main*

---

[1]Kenjiro Shoda (born February 25, 1902, died March 20, 1977) was a Japanese mathematician. Shoda was born in Tatebayashi, Gunma to a wealthy family. He was the second son of Teiichiro Shoda, the founder of Nisshin Flour, one of biggest companies in Japan. He was educated in Tokyo until he finished junior high school. He went to the National Eighth High School in Nagoya, today succeeded to Faculty of Liberal Arts of Nagoya University. After he finished the Eighth High School, he returned to Tokyo and studied mathematics at Imperial University of Tokyo. Shoda was supervised by Teiji Takagi, one of the best mathematicians in Japan at that time, and Takagi inspired Shoda to study algebra. Shoda graduated at Department of Mathematics, Faculty of Science at Tokyo University in 1925 and continued his graduate study under Takagi's supervision. In 1925, in his second year at Graduate School of Tokyo University, he got a scholarship which allowed him to study in Germany. With an interest in group theory, he went to Berlin to work with Issai Schur. After one year in Berlin, Shoda went to Göttingen to study with Emmy Noether. Noether's school brought a mathematical growth to him. In 1929 he returned to Japan. Soon afterwards, he began to write Abstract Algebra, his mathematical textbook in Japanese for advanced learners. It was published in 1932 and soon recognized as a significant work for mathematics in Japan. It became a standard textbook and was reprinted many times. In 1933 Shoda was appointed as professor in the Faculty of Science at Imperial Osaka University, which was founded in 1931 as the eighth Imperial University of Japan and hence the second one in the Kansai region, to promote industries in Osaka, therefore focusing on natural science, engineering and medicine in particular. The decades from the 1930s were a hard time for Japanese researchers. However, Shoda continued to apply himself to learning. After World War II, he was elected the first Chairman of the Mathematical Society of Japan in 1946. In this role he managed to reconstruct Japanese mathematics both theoretically and organizationally. Also, he was eager to attempt to keep the educational standard at Osaka University as its faculty staff. In this period, he published General Algebra, another textbook in Japanese. In 1949 Shoda was awarded the Japan Academy Prize in recognition of his fine achievements. Also that year he was elected the Dean of the Faculty of Science at Osaka University. In 1955 Shoda was appointed as President of Osaka University, a role in which he remained for six years. His achievements as President include foundations of two new faculties: the Faculty of Letters and the Faculty of Engineering Science, both based at Toyonaka, Osaka. The Faculty of Engineering Science was an ambitious attempt to synthesize two traditional disciplines: science and engineering. Some criticize the Faculty of Engineering Science as being nothing less than a duplicate of the Faculty of Engineering, while others recognize it as having helped to promote academic collaboration between multiple disciplines, including science, engineering and sometimes medical science. Shoda is remembered by the students and alumni of Osaka University as the founder of the Shoda Cup, which is given for the winning team of five people in an athletics contest. Shoda worried that most students were lacking in physical education and paid too little attention to it. With this Cup, he attempted to invoke interest for sporting activities among students. It succeeded and the Shoda Cup has been contested yearly by many students. When his term as President ended in 1961, Shoda left Osaka University but suddenly returned as a professor the Faculty of Engineering Science founded in this year and was appointed to its first Dean. After retirement from Osaka University, he worked still to improve the Japanese educational system in this field. He taught in Musashi University in Tokyo and became its President. In 1969 he was awarded the Order of Culture, and awarded the Grand Cordon of the Order of the Sacred Treasure in 1974. In 1977 Kenjiro Shoda died unexpectedly while driving with his family. He was posthumously raised to the second degree in the official order of precedence, and awarded the Grand Cordon of the Order of the Rising Sun. Kenjiro Shoda married twice. His first wife was Tami Hirayama, the daughter of astronomer Makoto

theorem on this topic?

Excercise 305 *If a meta-abelian has a faithful complex irreducible representation, then all maximal abelian subgroups have the same order (see previous exercise 304).*

Excercise 306 *Proof that direct products of ambivalent groups are ambivalent! Construct all involutions of a direct product of groups and determine the quantity with respect to the corresponding number of each factor! What is the consequence for the maximal tori of the group algebra of a direct product of groups? Determine this number for $S_n \times S_m$ with $n, m \in \mathbb{N}$! How can the formula be simplified for the case $n = m$?*

Excercise 307 *An ambivalent class of a group is a conjugacy class identical to its inverse class. Prove or falsify the following statements: If t is an involution of a group, then the conjugacy class containing t is ambivalent. The class containing the unit element is ambivalent, too. If g is an element of finite order not less than 3 and the conjugacy class containing g is finite and ambivalent, then this conjugacy class has even order. What are the ambivalent classes of groups of uneven order? Is there an ambivalent group of uneven order? A group is ambivalent if and only if all conjugacy classes are ambivalent.*

Excercise 308 *Determine the bound in corollary 24 for dihedral groups D_{2n}. Is there a group for which this bound is exact?*

Excercise 309 *Prove that the n-th root number function is a class function!*

Excercise 310 *Study the definition and meaning of the Frobenius-Schur indicators! What is exactly indicated by them?*

Excercise 311 *Let $n \in \mathbb{N}$ and G be a finite group. If n and $\mid G \mid$ are coprime than the n-th root number function is constant 1. Is the opposite statement true, too?*

Hirayama. He fathered one son and two daughters during this marriage. After the death of his first wife, Shoda married Sadako Ito, daughter of Eisaburo Ito, an engineering scientist and professor at Kyushu University. The remarried couple had one son. After his death, his family contributed a part of his legacy to some academic institutions including Osaka University. Osaka University used the money to make a small garden near to two of his former workplaces: the Faculty of Science and the Faculty of Engineering Science and named it after him. "Shoda Garden", a silent cozy space, is at the corner of main street of the campus, beside the building of the Cyber Media Center at Toyonaka, backed by dense bamboo woods. Sometimes people at Toyonaka hold their parties there, like a welcoming party for freshmen, or a barbecue just for fun and communication. Empress Michiko is one of his nieces.

Excercise 312 Determine the number of involutions for dihedral, semi-dihedral and quaternion groups. What is the relevance for the maximal tori in their complex group algebras?

Excercise 313 How can corollary 26 be extended to arbitrary groups?

Excercise 314 Determine the number of involutions for extra-special p-groups. What is the relevance for the maximal tori in its complex group algebra? Is the case $p = 2$ somehow special?

Excercise 315 Use corollary 25 for extra-special p-groups as well as for dihedral, semi-dihedral and quaternion groups. Determine under what terms these groups are nilpotent. Is the lower bound of this corollary greater than in corollary 24? Are one of these bounds exact for the groups mentioned here?

Excercise 316 Let K be a field and G a finite group such that KG is semisimple. Let L be a superfield and U a subfield of K. Then KG, LG and UG are semisimple and the dimensions of all Cartan subalgebras $(KG)^\circ$, $(LG)^\circ$ and $(UG)^\circ$ are identical.

Excercise 317 Let A be a K-algebra and $r, s \in \mathbb{N}$. The algebras $(A^{r \times r})^{s \times s}$ and $A^{rs \times rs}$ are isomorphic.

Excercise 318 Let K be a field, G a finite group and H a subgroup of G. Is KG semisimple, then KU is semisimple, too. Is the opposite implication true, too?

Excercise 319 Let K be a field, $n \in \mathbb{N}$ and A a finite-dimensional K-algebra. The K-algebras $A^{n \times n}$ and $K^{n \times n} \otimes_K A$ are isomorphic.

Excercise 320 Prove that the tensor product is a associative and commutative composition between algebras and is distributive with respect to the outer product of algebras.

Excercise 321 Let K be a field of characteristic p and G a finite p-group. Then $(KG)^\circ$ is nilpotent and the only Cartan subalgebra of $(KG)^\circ$ is KG. (Tip: theorem of Wallace; calculate $(g-1)^p$ for all $g \in G$)

Excercise 322 Let K be a field of characteristic q and G a finite p-group of order p^3. Determine the dimension of all Cartan subalgebras of $(KG)^\circ$ for the cases $p = q$ and for $p \neq q$! (Tip: extra-special groups and the previous exercise 321)

Excercise 323 Let $n \in \mathbb{N}$ with $n \geq 3$. Determine the conjugacy classes and the class number of SD_{2^n}! In addition, prove that the derived subgroup has the index 4!

Excercise 324 Let G be a finite group and K a field such that KG is semisimple. Then KG is separable. (Tip: use [74])

Excercise 325 Let G be a finite non-trivial group and K a field such that KG is semisimple. The lower bound $\sqrt{|G|}$ for the dimension of the Cartan subalgebras of $(KG)^\circ$ is never met.

Excercise 326 Determine the conjugacy classes and the class number for a dihedral group of order $2n$ for a natural number n.

Excercise 327 In a detailed proof determine the dimension of every Cartan subalgebra of $(KG)^\circ$ for the case that KG is semisimple and G is a generalized quaternion group of order $4n$ with $n \in \mathbb{N}_{\geq 2}$.

Excercise 328 Let G be a finite group, $n \in \mathbb{N}$, p a prime number, P a group of order p^3 and K a field. In the following cases analyze whether KG is semisimple and determine the dimension of every Cartan subalgebra of $(KG)^\circ$ in the following cases:

(i) $G = D_{2p}$

(ii) $G = D_{2p^n}$

(iii) $G = P^p$

(iv) $G = P \times D_{14}$

(v) $G = D_6$

(vi) $G = Q_8$

(vii) $G = D_{20}$

(viii) $G = Q_{12} \times SD_{2^n}$

(ix) $G = D_{2p} \times Q_{4n}$.

Excercise 329 Let K be a field of characteristic p, $n \in \mathbb{N}$, G a finite group and P a p-group. Determine the dimension of a maximal torus of $(KG)^\circ$ in the following cases:

(i) $G = D_{2n} \times P$, p does not divide $2n$

(ii) $G = Q_{4n} \times P$, p does not divide $4n$

(iii) $G = Q_{4p^n}$, $p \neq 2$

(iv) $G = SD_{2^n} \times P$, $p \neq 2$

(v) $G = Q_{2^n} \times P$, $p \neq 2$

(vi) $G = D_{2^n} \times P$, $p \neq 2$

(vii) $G = D_{2n} \times Q_{4n} \times P$, p does not divide $2n$

(viii) $G = D_{2^n} \times Q_{2^n} \times P$, $p \neq 2$

(ix) $G = D_{2^n} \times Q_{2^n} \times SD_{2^n} \times P$, $p \neq 2$.

Excercise 330 *Is it possible to derive the result of corollary 30 with the correspondent one for Frobenius groups?*

Excercise 331 *For the groups $S_3, A_3, S_4, A_4, S_5, A_5$ study whether there are normal Sylow subgroups and apply theorem 45 in theses cases! In addition calculate the degrees of the irreducible complex characters and use theorem 44!*

Excercise 332 *The square of the sum of finite many natural numbers is not less than the sum of squares of these numbers. Under what conditions are both values equal?*

Excercise 333 *What is the implication of the argumentation within corollary 51 for quaternion groups?*

Chapter 7

Invariants

Within chapter 7 we focus on the question whether the dimensions of the maximal tori and of the Cartan subalgebras are unique for associated Lie algebras of finite-dimensional associative unital algebras. For maximal tori we give a positive answer to this question for associative algebras with separable factor algebra by its nilradical by calculating this dimension explicitly. The answer for the Cartan subalgebras is positive, too. In characteristic zero we derive this result by using a classical result on Cartan subalgebras over algebraically closed fields. In the modular case we begin the analysis by proving the uniqueness for associated Lie algebras based on solvable finite-dimensional associative algebras, for separable associative algebras and for finite-dimensional associative algebras possessing a central nilradical. The general case is derived by using a result of Premet (which was later proven by Farnsteiner) for restricted Lie algebras over algebraically closed fields in positive characteristic and by using the result on the dimension for maximal tori. In general, the dimension of Cartan subalgebras can differ for restricted Lie algebras. By using a second approach we extend our theorem for the uniqueness of the dimension of Cartan subalgebras to the solvable and nilpotency class. For this, we prove that all maximal tori and Cartan subalgebras of Lie algebras associated to finite-dimensional associative algebras over an arbitrary algebraically closed field are conjugated. We demonstrates these three invariants – dimension, nilpotency and solvable class – by calculating them for group algebras based on dihedral and quaternion groups.

7.1 Dimension formula for maximal tori

Within the last chapter we have proven lemma 11 about maximal tori of factor algebras:

Lemma 12 *Let K be a field, A an associative finite-dimensional unitary K-algebra and N a nilpotent ideal of A. If T is a maximal torus of A, then $(T + N)/N$ is a maximal torus of A/N.*⋄

In addition, $(T+N)/N$ is isomorphic to $T/(T\cap N)$. T is a torus and contains no nilpotent elements different from zero. Hence, the intersection $T \cap N$ is exactly the null-space. As a consequence we obtain the following corollary:

Corollary 52 *Let K be a field, A an associative finite-dimensional unitary K-algebra. If T is a maximal torus of A, then $(T + rad(A))/rad(A)$ is a maximal torus of $A/rad(A)$ which is isomorphic to T. In particular, $dim_K(T) = dim_K((T + rad(A))/rad(A))$.* ⋄

By using this corollary we have reduced the question of a unique dimension for all maximal tori to semisimple algebras. Therefor we need to analyze the determination of maximal tori for direct products which is done within the next proposition:

Proposition 22 *Let K be a field, $n \in \mathbb{N}$ and $A_1, \ldots, A_n$ finite-dimensional associative K-algebras. The maximal tori of $A_1 \times \cdots \times A_n$ are exactly those associative subalgebras $C_1 \times \cdots \times C_n$, such that for every $i \in \underline{n}$ the set C_i is a maximal torus of A_i.*

Proof. By an induction argument the proposition needs to be proven only for two associative algebras A, B. Let C be a maximal torus of $A \times B$. We define $T := \{a \mid a \in A, \exists b \in B : (a, b) \in C\}$ and $S := \{b \mid b \in B, \exists a \in A : (a, b) \in C\}$. It is straightforward to prove that T resp. S is a commutative subalgebra of A resp. B. It is well-known that minimal polynomial of $(a;b) \in A \times B$ is the least common multiple of the minimal polynomials of a and b. Thus, S, T are tori. In particular, $T \times S$ is a torus of $A \times B$ containing by definition C as a subalgebra. As C is a maximal torus, we conclude $C = T \times S$. We have to prove that direct products of maximal tori are maximal tori. Let X resp. Y be a maximal torus of A resp. B. It is straightforward to prove that $X \times Y$ is a torus, too (Here we use again the statement about the minimal polynomial of $(a, b) \in A \times B$). Let C be a maximal torus of $A \times B$ containing $X \times Y$. As already proven, a torus T of A and S of B exist such that $C = T \times S$ is valid. Hence, $X \times Y \subseteq T \times S$ is true and we conclude $X \subseteq T$ and $Y \subseteq S$. By the maximality of X, Y we derive the maximality of $X \times Y$. ⋄

Proposition 22 and corollary 52 reduce the question of a unique dimension of maximal tori to simple associative algebras. For these algebras we have already calculated this dimension in corollary 3 if the algebra is separable:

Corollary 7 *Let A be a simple finite-dimensional associative separable K-algebra. The Cartan subalgebras of A° are identical with the maximal tori of A, and they are of dimension $dim_K(Z(A)) \cdot ind_{Z(A)}(A)$.* ⋄

In summarizing, proposition 22 and corollaries 52 and 7 let us derive the following theorem about the dimension of the maximal tori:

Theorem 46 *Let A be a finite-dimensional associative unital algebra with separable factor algebra by its nilradical, $r, n_1, \ldots, n_r \in \mathbb{N}$ and $D_1, \ldots, D_r$ division algebras such that $A/rad(A)$ is isomorphic to $\bigoplus_{i=1}^{r} D_i^{n_i \times n_i}$. The maximal tori of A are isomorphic to the maximal tori of $A/rad(A)$, and they are of the unique dimension*

$$\sum_{i=1}^{r} dim_K(Z(D_i)) \cdot ind_{Z(D_i)}(D_i^{n_i \times n_i}) = \sum_{i=1}^{r} n_i \cdot dim_K(Z(D_i)) \cdot ind_{Z(D_i)}(D_i). \diamond$$

Let A be a finite-dimensional associative unital algebra with separable factor algebra by its nilradical. The unique dimension of all maximal tori is denoted by $mto(A)$. By theorem 46 the identity

$$mto(A) = mto(A/rad(A)) = \sum_{i=1}^{r} n_i \cdot dim_K(Z(D_i)) \cdot ind_{Z(D_i)}(D_i)$$

is valid.

If A is solvable, then we have proven that the radical complements are exactly the maximal tori. Indeed, the dimension formula for fields D_i and $n_i = 1$ for all $i \in \underline{r}$ is reduced to

$$mto(A) = mto(A/rad(A)) = \sum_{i=1}^{r} n_i \cdot dim_K(Z(D_i)) \cdot ind_{Z(D_i)}(D_i) =$$

$$\sum_{i=1}^{r} dim_K(D_i) = dim_K(A/rad(A)).$$

For modular group algebras we have proven in lemma 10 that the factor algebra by its nilradical is isomorphic to a direct sum of full matrix algebras over fields. If D_i is a field for all $i \in \underline{r}$, then the formula is reduced to

$$mto(A) = mto(A/rad(A)) = \sum_{i=1}^{r} n_i \cdot dim_K(D_i).$$

For an algebraically closed field K all division algebras are identical to the base field K. For those algebras the dimension formula is reduced to

$$mto(A) = mto(A/rad(A)) = \sum_{i=1}^{r} n_i.$$

Within the exercises the reader may calculate the formula for other classes of associative algebras.

7.2 Cartan subalgebras

Let L be a Lie algebra. If the dimension of all Cartan subalgebras is identical, then we denote this common dimension by $cta(L)$.

In chapter 5 we have proven for finite-dimensional associative separable algebras A that the maximal tori of A coincide with the Cartan subalgebras of A°. Thus – by using theorem 46 – we conclude:

Theorem 47 *Let K be a field and A an associative separable K-algebra. The dimension of all Cartan subalgebras is identical to:*

$$cta(A^\circ) = mto(A) = mto(A/rad(A)) = \sum_{i=1}^{r} n_i \cdot dim_K(Z(D_i)) \cdot ind_{Z(D_i)}(D_i).\diamond$$

A special case of a separable algebra is a semisimple group algebra. Within chapter 6, theorem 44 we have proven that the dimension is unchanged for all fields such that the group algebra semisimple. Therefor we derive:

Corollary 53 *Let G be a finite group, K a field such that KG is semisimple and $\chi_1, \chi_2, \cdots, \chi_{k(G)}$ the complex irreducible characters of G. Every Cartan subalgebra H of $(KG)^\circ$ possesses the dimension*

$$cta((KG)^\circ) = mto(KG) = \sum_{i=1}^{k(G)} \deg(\chi_i).\diamond$$

In chapter 5, theorem 24 we haven proven that for a finite-dimensional associative unitary solvable K-algebra with separable factor algebra by its nilradical all Cartan subalgebras are conjugated and that they are centralizers of the maximal tori (which are exactly the radical complements). If T is a maximal tori of A, then $C_A(T) = T \oplus C_{rad(A)}(T)$ is valid. By using theorem 46 we know surplus that all maximal tori have the same dimension, too. Thus, we conclude:

Theorem 48 *Let A be a finite-dimensional associative unitary solvable K-algebra with separable factor algebra by its nilradical. Then all Cartan subalgebras of A° have the same dimension. If C is a Cartan subalgebra of A° based on a maximal tori T, then $cta(A^\circ) = dim_K(A/rad(A)) + dim_K(C_{rad(A)}(T))$ and $mto(A) = dim_K(A/rad(A))$ are valid. In particular, if S is another maximal tori, then we derive $dim_K(C_{rad(A)}(T)) = dim_K(C_{rad(A)}(S)).\diamond$*

Our next aim is to prove the uniqueness of the dimension for Cartan subalgebras within theorem 48 without assuming that factor algebra by the nilradical is separable and that the algebra is unitary.

Theorem 49 *Let K be a field and A a finite-dimensional associative solvable K-algebra. Then all Cartan subalgebras of A° have the same dimension.*

Proof. Within section 5.10.3 we have proven that the Cartan subalgebras of $(A^K)^\circ$ are exactly the subalgebras (C, K) for which C is a Cartan subalgebra of A°. Therefor the dimension of the Cartan subalgebras is unique for $(A, K)^\circ$ if and only if its unique for A°. In this case $cta((A, K)^\circ) = 1 + cta(A^\circ)$ is valid. Thus we can assume that A is unitary.

Let L be an algebraically closed superfield of K and C a Cartan subalgebra of A°. Within section 5.10.5 we have mentioned that $C \otimes L$ is a Cartan subalgebra of $(A \otimes L)^\circ$ with $dim_K(C) = dim_L(C \otimes L)$. Therefor we have to prove that the dimension of the Cartan subalgebras of $(A \otimes L)^\circ$ is unique. For this, it is sufficient to derive that $A \otimes L$ is solvable. If $A \otimes L$ is solvable as an L-algebra, then its factor algebra by the nilradical is isomorphic to a direct product of full matrix algebras over L because L is algebraically closed. By theorem 48 the proof is finished.

We prove now that $A \otimes_K L$ is solvable. By using remark 8 the subalgebra $(A \otimes_K L) \circ (A \otimes_K L)$ is contained in $(A \circ A) \otimes_K L$. A is solvable and hence $A \circ A$ is contained in $rad(A)$. We conclude that the derivation of $(A \otimes_K L)^\circ$ is contained in $rad(A) \otimes_K L$. Again by using remark 8 this subalgebra is nilpotent and therefor contained in the nilradical of $A \otimes L$.⋄

Within section 5.9.2 we haven proven for an associative finite-dimensional unitary K-algebra A with separable factor algebra by its nilradical that the Cartan subalgebras are of the form $T \oplus C_{rad(A)}(T)$, T a maximal torus of a radical complement. By theorem 46 we know that the dimensions of the maximal tori are identically. Hence the dimension of all Cartan subalgebras of A° is unique if and only if $C_{rad(A)}(T)$ is unique for all maximal tori of A. This is for example valid if $C_{rad(A)}(T) = rad(A)$ is true for all maximal tori T of A. This identity is valid if the radical is central. Thus, we conclude the following theorem:

Theorem 50 *Let K be a field and A a finite-dimensional associative unitary K-algebra with separable factor algebra by its nilradical. If the nilradical is central, then all Cartan subalgebras of A° have the same dimension $cta(A^\circ) = mto(A) + dim_K(rad(A))$. All Cartan subalgebras of A° are of the form $T \oplus rad(A)$ for a maximal torus T of A.⋄*

Let G be a finite group and K a field of characteristic p. We have used a result of Karpilovsky (in [33], chapter 3, corollary 1.18) that the factor algebra of the nilradical of KG is separable for every finite group and arbitrary field. For modular group algebras the condition of possessing a central nilradical is well-analyzed by Wallace and Siegel (see e.g. [33], chapter 3, theorem 13.4): p is not dividing the order of G or G is abelian or for every $P \in Syl_p(G)$ the group $G'P$ is a Frobenius group with complement P and kernel G'. In the first case the group algebra is semisimple, and this situation is already analyzed within chapter 6. The second condition is equivalent to the commutativity of the group algebra, and thus its Lie nilpotent. The more interesting case is the third one. Within the same chapter it is proven in theorem 15.2 that the condition of possessing a central nilradical is equivalent for $p \neq 2$ to the commutativity of the nilradical of the group algebra. We conclude by using theorem 50 and the theorem of Wedderburn-Malcev:

Theorem 51 *Let G be a finite non-abelian group and K a field of characteristic p such that p divides the order of G. The nilradical is central if and only if for every $P \in Syl_p(G)$ the group $G'P$ is a Frobenius group with complement P and kernel G'. For $p \neq 2$ this condition is equivalent to the commutativity of the nilradical. In particular, KG possesses exactly one radical complement. If the nilradical is central, then all Cartan subalgebras of $(KG)^\circ$ have the same dimension $cta((KG)^\circ) = mto(KG) + \dim_K(rad(KG))$. All Cartan subalgebras of $(KG)^\circ$ are of the form $T \oplus rad(KG)$ for an maximal torus T of KG.*⋄

Within the next example we analyze the symmetric group S_3 for an arbitrary field K whether the dimension of the Cartan subalgebras of $(KS_3)^\circ$ is unique. The reader might proof this example in details in exercise 372.

Example 10 Let $G := S_3$ and K be a field.

Case 1: In this case the group algebra is semisimple. By our results of chapter 6 the dimension of the Cartan subalgebras of $(KG)^\circ$ is unique and exactly the unique dimension of all maximal tori of KG. G is ambivalent, and hence this dimension is exactly $1 + t(G) = 1 + t(S_3) = 4$.

Case 2: In this case the characteristic of the field is 3. By our results for solvable group algebras in section 5.5 we know that the dimension is unique. S_3 is isomorphic to $D_{2 \cdot 3}$, and hence the dimension is exactly $3 + 1 = 4$.

Case 3: In this case the characteristic of the field is 2. It is well-known that S_3 is a Frobenius group with kernel A_3. Each of the 2-Sylow subgroups is a complement of A_3. We can apply theorem 51 and conclude that the nilradical is central. Let A_3 generated by an element a of order 3 and b be an involution of S_3. The element $a + a^2$ is central (its a conjugacy class sum) in KG and $(a+a^2)^2 = a^2 + a^3 + a^3 + a^4 = a^2 + a^4 = a + a^2$ is valid. The center of KG is 3-dimensional and $\langle 1, a+a^2 \rangle_K$ is a torus of KG. Thus, the radical is one-dimensional and the radical complement is 5-dimensional. KG is not solvable and possesses exactly one radical complement. (This statement is nearly the same as for Lie nilpotent associative algebras proven in section 5.6.) It is straightforward to prove that the complement is isomorphic to

$K \times K^{2 \times 2}$. The maximal tori of the complement are of dimension 3. Hence all Cartan subalgebras of $(KG)^{\circ}$ are again of dimension 4. At the end of this section we will extend this example to dihedral groups D_{2p^n} for an uneven prime number p. Within this example for every choice of the field the dimension of a Cartan subalgebra is unchanged. This is not true in general which the next example demonstrates.◊

Example 11 Let p be a prime number, G a finite non-abelian p-group and K a field.

Case 1: In this case the group algebra is semisimple. Hence by theorem 44 the dimension of each Cartan subalgebra is exactly the sum of the degrees of all irreducible complex characters. This sum is smaller than the group order because G is not abelian and the sum of all squares of the degrees of all irreducible characters is exactly the order of G. E.g. this sum is $2p^2 - p$ (see corollary 31) for an extra-special p-group G of order p^3. This sum is smaller than p^3.

Case 2: In this case the characteristic of the field is p and thus the group algebra is Lie nilpotent due to a theorem of Wallace. Thus, the dimension of the only Cartan subalgebra KG is the order of G.

This example demonstrates that the dimension of a Cartan subalgebra of $(KG)^{\circ}$ for a fixed finite group and a field K can be varied for different choices of the field K because the corresponding group algebras can have a different structure. But, if the field K is fixed, too, then the dimension of all Cartan subalgebras is unique. We will prove this theorem in this section for a Lie algebra associated to an arbitrary associative finite-dimensional algebra by using two different approaches.◊

We begin to prove the uniqueness by using our first approach. The proof is based on the main theorem about the uniqueness of the dimension of maximal tori in section 7.1 (theorem 46) and a theorem of Premet (1987), which was proven again by Farnsteiner in 2004. I want to say thank you to Mr. Farnsteiner for providing this result to me (see [50] and [15]):

Lemma 13 *(Premet, 1987, Farnsteiner, 2004) Let K be an algebraically closed field with $char(K) = p \in \mathbb{P}$ and L a finite-dimensional restricted K-Lie algebra. If T is a torus of maximal dimension, then $C_L(T)$ is a Cartan subalgebra of minimal dimension (= the rank of L).*

In particular, if all maximal tori possess the same dimension, then all Cartan subalgebras possess the same dimension, too.◊

Theorem 52 *Let K be a field and A an associative finite-dimensional unitary K-algebra. The Cartan subalgebras of A° exist and all possess the unique dimension $cta(A^\circ)$.*

If $A/rad(A)$ is separable, then for every maximal tori T of A the identity $cta(A^\circ) = mto(A^\circ) + dim(C_{rad(A)}(T))$ is valid. In particular, $dim_K(C_{rad(A)}(T))$ is unique for every maximal torus T of A.

Proof. The existence of Cartan subalgebras is proven within theorem 20. Let L be an algebraically closed superfield of K and H a Cartan subalgebra of A°. Then (see e.g. [29]) $H \otimes L$ is a Cartan subalgebra of $(A \otimes L)^\circ$. We have to prove that all Cartan subalgebras of $(A \otimes L)^\circ$ possess the same dimension. If the field is of characteristic zero, then all Cartan subalgebras are conjugated by a classical result on Cartan subalgebras. Hence, all Cartan subalgebras possess the same dimension. Let L be a field with $char(L) = p \in \mathbb{P}$. $(A \otimes L)^\circ$ is a restricted Lie algebra by the ordinary p-power mapping and all maximal tori of this Lie algebra coincide with the ones of $A \otimes L$ by proposition 10. L is algebraically closed and hence the factor algebra of $A \otimes L$ by its nilradical is isomorphic to a product of full matrix algebras over L (by Wedderburn) which is a separable K-algebra. By using theorem 46 all maximal tori of $A \otimes L$ are of the same dimension $mto(A \otimes L)$. We finish the proof for the uniqueness of the dimension by applying the theorem of Premt/Farsteiner 13 to $(A \otimes L)^\circ$. The add-on is valid by using the theorems 42 and 46: if T is a maximal torus of A, then $C_A(T) = T \oplus C_{rad(A)}(T)$ is valid. The dimensions of $C_A(T)$ (Cartan subalgebra) and T (maximal torus) are unique, hence the dimension of $C_{rad(A)}(T)$ is unique, too.⋄

We have used a result of Karpilovsky (in [33] and in [34]) that the factor algebra of the nilradical in KG is separable for every finite group and arbitrary field. Therefor we derive by theorem 52 for group algebras:

Theorem 53 *Let K be a field and G a finite group. Cartan subalgebras of $(KG)^\circ$ exist and all possess the unique dimension $cta((KG)^\circ)$.*

For every maximal tori T of KG the identity $cta((KG)^\circ) = mto((KG)^\circ) + dim_K(C_{rad(A)}(T))$ is valid. In particular, $dim_K(C_{rad(A)}(T))$ is unique for every maximal torus T of KG.⋄

The second approach provides a deeper insight in the toral structure of Lie algebras associated to associative algebras over algebraically closed fields. In fact, we will prove that all maximal tori are conjugated. Hence, the Cartan subalgebras are conjugated, too. As a consequence we derive that the dimension but also the class of nilpotency and the solvable class is unique for all Cartan subalgebras.

Another consequence is derived for the weights and the weight decomposition with respect to the Cartan subalgebras. Let L be a K-Lie algebra and C a subalgebra of L. A function $\alpha : C \longrightarrow K$ is called a weight if the subspace $L_\alpha := \{x \mid x \in L, \forall c \in C \ \exists n \in \mathbb{N} : x(ad(y) - \alpha(y)id)^n = 0\}$ is not zero. In this case L_α is called the weight space of α. The weight spaces with respect to a Cartan subalgebra C play a grave role within the theory of Lie algebras. Cartan subalgebras are weight spaces with respect to the function $\alpha : C \longrightarrow K, c \mapsto 0$. In [40] it is proven that a Lie algebra possesses a decomposition into a finite direct sum of weight spaces if for all $x \in L$ the characteristic polynomial of $ad(l)$ is a product of polynomials of degree 1 for all $l \in L$. In particular, such a decomposition exists for a Lie algebra over an algebraically closed field.

Lemma 14 *Let K be an algebraically closed field and A a finite-dimensional associative unitary K-algebra. All maximal tori of A are conjugated. In particular, all Cartan subalgebras of A° are conjugated, too.*

Add-on: The decomposition of A° into weight spaces with respect to a Cartan subalgebra exists and is unique up to isomorphism and does not depend on the choice of the Cartan subalgebra. Two weight decompositions based on conjugated Cartan subalgebras possess the same weight functions and isomorphic weight spaces. In particular, the weights and the dimensions of the weight spaces are invariants for A°.

Proof. The factor algebra by the nilradical of A is isomorphic to a direct product of full matrix algebras over L because L is algebraically closed. In particular, this factor algebra is separable. By using the theorem of Wedderburn-Malcev let T be a unital subalgebra of A, $I_1, \cdots, I_r$ simple ideals of T and $n_1, \cdots, n_r \in \mathbb{N}$ such that $T = I_1 \oplus \cdots \oplus I_r$ and $I_i \cong L^{n_i \times n_i}$ are valid for all $1 \leq i \leq r$. The Cartan subalgebras are the centralizers of the maximal tori of A (see theorem 20). We prove that all maximal tori are conjugated. Thus, all Cartan subalgebras are conjugated, too, and the theorem is proven. For all $1 \leq i \leq r$ let e_i be the unit of I_i. T is a unital subalgebra of A and $1_A = 1_T = e_1 + \cdots + e_r$ is valid (see e.g. [74], chapter 1). Let X be a maximal torus of A. Then X can be conjugated into T (by a generalized version of the Wedderburn-Malcev theorem, see e.g [74]). T is separable and hence maximal tori and Cartan subalgebras are identical (see theorem 42) and they are the direct sum of the Cartan subalgebras of the simple components of T (see theorem 35). Hence for all $1 \leq i \leq r$ there exists a Cartan subalgebra X_i of I_i with $X = X_1 + \cdots + X_r$. By a theorem of Jacobson (see e.g. [29]) the Cartan subalgebras of full matrix algebras over algebraically closed fields are exactly the conjugates of the subalgebra of diagonal matrices. For all $1 \leq i \leq r$ let D_i be a subalgebra of I_i isomorphic to the subalgebra of diagonal matrices of $L^{n_i \times n_i}$. We conclude that for all $1 \leq i \leq r$ there exists a unit g_i in I_i such that $X_i^{g_i} = D_i$ is valid. The element $g := g_1 + \cdots + g_r$ is a unit of A: let $h_i \in I_i$ with $g_i \cdot h_i = e_i = h_i \cdot g_i$ for all $1 \leq i \leq r$. Then $(h_1 + \cdots + h_r) \cdot g = g \cdot (h_1 + \cdots + h_r) = e_1 + \cdots + e_r = 1_A$ because the ideals $I_1, \cdots, I_r$ are direct. Again by using their directness we conclude $X^g = X_1^{g_1} + \cdots + X_r^{g_r} = D_1 + \cdots + D_r$. The add-on is a consequence of the fact that all Cartan subalgebras are conjugated.⋄

The lemma is used to derive three invariants of Cartan subalgebras. Let K be a field, A a finite-dimensional associative unitary K-algebra, L an algebraically closed superfield of K and H a Cartan subalgebra of $A°$. By a remark of Jacobson in [29] the subalgebra $H \otimes L$ is a Cartan subalgebra of $(A \otimes L)°$ which possesses the same dimension and (by using remark 8) also the same class of nilpotency and solvable class as H. Within $(A \otimes L)°$ these three values are unique, and hence they are unique in $A°$, too. We conclude:

Main theorem 2 *Let K be a field and A a finite-dimensional associative unitary K-algebra. Cartan subalgebras of $A°$ exist and they are of*

(i) unique dimension $cta(A°)$,

(ii) unique nilpotency class $ctn(A°)$ and of

(iii) unique solvable class $cts(A°)$.⋄

In section 5.10. we have analyzed the determination of one explicit Cartan subalgebra for certain algebra constructions. In the next corollary we use this analysis and the main theorem 2 to derive some results about the dimension, nilpotency and solvable class of Cartan subalgebras for these algebra constructions. In particular, we extend the main theorem to non-unitary algebras:

Corollary 54 *Let K be a field, L a superfield of K, $n \in \mathbb{N}$, C a resp. A, B two finite-dimensional associative non-unitary resp. unitary K-algebras, H resp. E a Cartan subalgebra of $A°$ resp $B°$ and I an ideal of A. The following statements are valid:*

(i) $cta((A^K)°) = cta(A°)+1$, $ctn((A^K)°) = ctn(A°)$, $cts((A^K)°) = cta(A°)$

(ii) $cta((A \otimes L)°) = cta(A°)$, $ctn((A \otimes L)°) = ctn(A°)$, $cts((A \otimes L)°) = cts(A°)$

(iii) $cta((A \times B)°) = cta(A°)+cta(B°)$, $cts((A \times B)°) = max\{cts(A°), cts(B°)\}$, $ctn((A \times B)°) = max\{ctn(A°), ctn(B°)\}$

(iv) $cta((A^{n \times n})°) = n \cdot cta(A°)$, $ctn((A^{n \times n})°) = ctn(A°)$, $cts((A^{n \times n})°) = cts(A°)$

(v) If $dim_K(A) < |K|$, then $cta((A/I)^\circ) \leq cta(A^\circ)$, $ctn((A/I)^\circ) \leq ctn(A^\circ)$ and $cts((A/I)^\circ) \leq cts(A^\circ)$ are valid. In addition, $cta(A^\circ) = cta((A/I)^\circ) + dim_K(H \cap I)$ is true and $dim_K(H \cap I)$ is independent of the choice of H.

(vi) $cta((A \otimes B)^\circ) = cta(A^\circ) \cdot cta(B^\circ)$.

Proof. By using main theorem 2 it is sufficient to prove the identities for one Cartan subalgebra.

ad(i): By the results of section 5.10.3 we extend the main theorem 2 to non-unitary K-algebras. In addition, we have proven in that section that (H, K) is a Cartan subalgebra of $(A^K)^\circ$. $(H, K) = (H, 0) + (0, K)$ is valid and $(0, K)$ is central. Thus, the nilpotency and solvable class of H and (H, K) are identical.

ad(ii): By a remark of Jacobson in [29] the subalgebra $H \otimes L$ is a Cartan subalgebra of $(A \otimes L)^\circ$ which have the same dimension and – by using remark 8 – also the same class of nilpotency and solvability as H.

ad(iii): By theorem 35 the subalgebra $H \times E$ is a Cartan subalgebra of $A \times B$, and the dimension is the sum of the dimensions of H and E. Lie products of $H \times E$ are determined within the factors. Thus the nilpotency and solvable class is the maximum of the nilpotency and solvable class of H and E.

ad(iv): By proposition 21 the subalgebra of diagonal matrices over H in $A^{n \times n}$ is a Cartan subalgebra of $(A^{n \times n})^\circ$. The dimension of this subalgebra is $n \cdot dim_K(H)$. Lie products within this subalgebra are diagonal matrices such that every entry is a Lie product of A° of the same length and form. Thus, the nilpotency and solvable class of H is the same as for the subalgebra of diagonal matrices over H.

ad(v): By corollary 20 the subalgebra $(H + I)/I$ is a Cartan subalgebra of $(A/I)^\circ$. This subalgebra is isomorphic to $H/(H \cap I)$. Thus, the dimension, nilpotency and solvable class of $H/(H \cap I)$ is not greater than the corresponding dimension and classes of H. The add-on is deducible by the dimension formula $dim_K(H/(H \cap I)) = dim_K(H) - dim_K(H \cap I)$ because by using the main theorem 2 we conclude $cta((A/I)^\circ) = dim_K(H/(H \cap I))$ and $cta(A^\circ) = dim_K(H)$.

ad(vi): Let M be an algebraically closed superfield of K. We focus on the base field extension $(A \otimes_K B) \otimes M$. By part (ii) the dimension of the Cartan subalgebras do not change by using a base field extension. The factor algebras by the nilradical of the M-algebras $A \otimes_K M$ and $B \otimes_K M$ are separable because M is algebraically closed. For every field extension M (here we do not need that M is algebraically closed) there is a canonical isomorphism between $(A \otimes_K B) \otimes M$ and $(A \otimes_K M) \otimes_M (B \otimes_K M)$ (see e.g. [49]). By using theorem 43 the dimension of a special Cartan subalgebra of $(A \otimes_K M) \otimes_M (B \otimes_K M)$ is the product of the dimension of a special Cartan subalgebra of $A \otimes_K M$ and $B \otimes_K M$.⋄

Our next aim is to extend example 10 to the groups D_{2p^n}, D_{4p^n} and Q_{4p^n}. For group algebras the following observation is valid:

Proposition 23 *Let G be a finite group and K, L fields with $char(K) = char(L)$. Then $cta((KG)^\circ) = cta((KG)^\circ)$, $ctn((KG)^\circ) = ctn((KG)^\circ)$ and $cts((KG)^\circ) = cts((KG)^\circ)$ are valid.*

Proof. Both fields possess isomorphic prime fields $P(K)$ and $P(L)$ because they have the same characteristic. Thus, the corresponding group algebras $P(K)G$ and $P(L)G$ are isomorphic and $cta((P(K)G)^\circ) = cta((P(L)G)^\circ)$ is valid. It is well-known that $P(K)G \otimes K$ and KG are isomorphic, and by corollary 54 we derive $cta((P(K)G)^\circ) = cta((P(K)G \otimes K)^\circ)$. Hence we conclude $cta((P(K)G)^\circ) = cta((KG)^\circ$. With an analogue argument we derive $cta((P(L)G)^\circ) = cta((KL)^\circ$, and the proof is finished for cta. An analogue argument is valid for ctn and cts, too (The nilpotency class and solvable class do not change under base field extensions).⋄

Example 12 *(dihedral groups of order $2p^n$)* Let n be a natural number, p an uneven prime number, $G := D_{2p^n}$ and K a field.

Case 1: Let KG be a semisimple which is valid for $char(K) \neq 2$ and $char(K) \neq p$. By theorem 44 the maximal tori are exactly the Cartan subalgebras of $(KG)^\circ$. Hence we deduce $ctn((KG)^\circ) = cts((KG)^\circ) = 1$ and $mto(KG) = cta((KG)^\circ)$. Within corollary 32 we have determined the dimension of the Cartan subalgebras of $(KG)^\circ$: $cta((KG)^\circ) = p^n + 1$.

Case 2: Let $char(K) = p$. In this case the group algebra is solvable as described within section 5.5.4. The maximal tori are exactly the radical complements and all are conjugated by the Wedderburn-Malcev theorem. Their dimension is $mto(KG) = 2$. The Cartan subalgebras are the centralizers of the radical complements and all are conjugated, too. Their dimension is $cta((KG)^\circ) = p^n + 1$. An explicit construction is given within section 5.5.4 for one Cartan subalgebra. We use this construction to determine that all Cartan subalgebras are abelian: $ctn((KG)^\circ) = cts((KG)^\circ) = 1$. Let $a, b \in G$, $o(a) = p^n$ and $o(b) = 2$ such that $a^b = a^{-1}$ and $G = \langle a, b \rangle$. If $H := \langle b \rangle$, then KH is a maximal torus of KG and $C_{KG}(KH)$ is a Cartan subalgebra of $(KG)^\circ$. Within section 5.5.4 we have proven that $C_{KG}(KH)$ is linear spanned by the orbit-sums of the action of H on G. For every $i \in \mathbb{N}$

the elements a^i and a^{-i} are conjugates in G with respect to $b \in H$. The G-conjugacy classes b^G, $(ab)^G$ decompose under the action of H – and then added together in KG – into $1, b, z_j b$ such that z_j is central in KG. z_j is of the form $a^j + a^{-j}$ for some $j \in \mathbb{N}$. We conclude that the elements of a basis of $C_{KG}(KH)$ are $1, b, z_j, z_j b$ for several $j \in \mathbb{N}$. By using this result it is straightforward to prove that $C_{KG}(KH)$ is commutative.

Case 3: Let $char(K) = 2$. In this case we can apply theorem 51. $G'P$ is a Frobenius group for all 2-Sylow subgroups P of G. Hence for all maximal tori T of KG we derive $C_{KG}(T) = rad(KG) \oplus T$ and $rad(KG)$ is central. Thus all Cartan subalgebras are abelian: $ctn((KG)^\circ) = cts((KG)^\circ) = 1$. The nilradical of KG is determined within [33] to be the one-dimensional subspace $\langle \overline{G} \rangle_K$. Therefor we conclude $mto(KG) + 1 = cta((KG)^\circ)$. We prove that $mto(KG) = p^n$ and $cta((KG)^\circ) = p^n + 1$ are valid. In particular, for all fields K the dimension of the Cartan subalgebras of $(KG)^\circ$ is identical. By using proposition 23 it is sufficient to analyze the dimension of the Cartan subalgebra for a finite field F_q of characteristic 2. N. Makhijani, R. K. Sharma and J. B. Srivastava have analyzed the group algebra of dihedral groups for F_q in their articles [41] and [42]. For every divisor m of p^n there exist a natural number e_m (the exact definition of e_m is not needed in our context) such that the factor algebra $KG/rad(KG)$ is isomorphic to $A := F_q \oplus \bigoplus_{m|p^n, m>1} ((F_{q^{e_m}})^{2 \times 2})^{\frac{\phi(m)}{2e_m}}$. We prove that $mto(A) = p^n$ is valid. By using theorem 53 we conclude $mto(KG) = mto(A) = cta(A^\circ)$. We apply main theorem 2, corollaries 54 and 3 and theorem 36 to determine this value, and we conclude $cta(A^\circ) = 1 + \sum_{m|p^n, m>1} \frac{\phi(m)}{2e_m} \cdot 2 \cdot dim_{F_q}(F_{q^{e_m}})$. Based on the theory of finite fields the identity $dim_{F_q}(F_{q^{e_m}}) = e_m$ is valid. Hence we derive $cta(A^\circ) = 1 + \sum_{m|p^n, m>1} \phi(m)$. The sum is exactly $p^n - 1$ (see e.g. exercise 199). For all fields K the solvable and nilpotency class of the Cartan subalgebras of $(KG)^\circ$ are identical. ◇

Example 13 *(dihedral groups of order $4p^n$)* Let n be a natural number, p an uneven prime number, $G := D_{4p^n}$ and K a field.

Case 1: Let KG be a semisimple which is valid for $char(K) \neq 2$ and $char(K) \neq p$. By theorem 44 the maximal tori are exactly the Cartan subalgebras of $(KG)^\circ$. Hence we deduce $ctn((KG)^\circ) = cts((KG)^\circ) = 1$ and $mto(KG) = cta((KG)^\circ)$. Within corollary 32 we have determined the dimension of the Cartan subalgebras of $(KG)^\circ$: $cta((KG)^\circ) = p^n + 2$.

Case 2: Let $char(K) = p$. In this case the group algebra is solvable as described within section 5.5.4. The maximal tori are exactly the radical complements which are all conjugated by the Wedderburn-Malcev theorem.

Their dimension is $mto(KG) = 4$. The Cartan subalgebras are the centralizers of the radical complements and all are conjugated, too. Their dimension is $cta((KG)^\circ) = p^n + 2$. An explicit construction is given within section 5.5.4 for one Cartan subalgebra. We use this construction to determine that all Cartan subalgebras are abelian: $ctn((KG)^\circ) = cts((KG)^\circ) = 1$. Let $a, b \in G$, $o(a) = 2 \cdot p^n$ and $o(b) = 2$ such that $a^b = a^{-1}$ and $G = \langle a, b \rangle$. If $z := a^{p^n} b$ and $H := \{1, b, z, bz\}$, then z is central, KH is a maximal torus of KG and $C_{KG}(KH)$ is a Cartan subalgebra of $(KG)^\circ$. Within section 5.5.4 we have proven that $C_{KG}(KH)$ is linear spanned by the orbit-sums of the action of H on G. For every $i \in \mathbb{N}$ the elements a^i and a^{-i} as well as za^i and za^{-i} are conjugates in G with respect to $b \in H$. The other G-conjugacy classes decompose under the action of H – and then added together in KG – into $b, bz, z_j b, zz_j b$ such that z_j is central in KG. z_j is of the form $a^j + a^{-j}$ for some $j \in \mathbb{N}$. We conclude that the elements of a basis of $C_{KG}(KH)$ are $1, z, b, bz, z_j, z_j b, zz_j b$ for several $j \in \mathbb{N}$. By using this result it is straightforward to prove that $C_{KG}(KH)$ is commutative.

<u>Case 3:</u> Let $char(K) = 2$. By using proposition 23 it is sufficient to analyze the dimension, the solvable and nilpotency class of the Cartan subalgebras for a finite field F_q of characteristic 2. N. Makhijani, R. K. Sharma and J. B. Srivastava have analyzed the group algebra of dihedral groups for F_q in their articles [41] and [42]. For every divisor m of p^n there exist a natural number e_m (the exact definition of e_m is not needed in our context) such that the factor algebra $KG/rad(KG)$ is isomorphic to $A := F_q \oplus \bigoplus_{m|p^n, m>1} ((F_{q^{e_m}})^{2 \times 2})^{\frac{\phi(m)}{2e_m}}$. As proven in example 12 we derive $mto(A) = mto(KG) = p^n$. A maximal torus of KG is $K\langle a^2 \rangle$. Thus we have to analyze the Cartan subalgebra $C_{KG}(\langle a^2 \rangle)$. For this the action of $\langle a^2 \rangle$ on G is to be determined. Every element of $\langle a \rangle$ is stable under the action of $\langle a^2 \rangle$. The G-conjugacy classes b^G and $(ab)^G$ are also classes under the action of $\langle a^2 \rangle$ because the centralizer of b resp. ab in G is $\langle b, z \rangle$ resp. $\langle ab, z \rangle$. They possess trivial intersection with $\langle a^2 \rangle$. Thus, $C_{KG}(\langle a^2 \rangle)$ is linear spanned by $\langle a \rangle$ and the central elements $\overline{b^G}$ and $\overline{(ab)^G}$. In particular, the centralizer is commutative and of dimension $p^n + 2$: for all fields K the dimension of the Cartan subalgebras of $(KG)^\circ$ is identical as well as the solvable and nilpotency class.◊

Example 14 *(quaternion groups of order $4p^n$)* Let n be a natural number, p an uneven prime number, $G := Q_{4p^n}$ and K be a field.

<u>Case 1:</u> Let KG be a semisimple which is valid for $char(K) \neq 2$ and $char(K) \neq p$. By theorem 44 the maximal tori are exactly the Cartan subalgebras of $(KG)^\circ$. Hence we deduce $ctn((KG)^\circ) = cts((KG)^\circ) = 1$ and $mto(KG) = cta((KG)^\circ)$. Within corollary 33 we have determined the

dimension of the Cartan subalgebras of $(KG)^\circ$: $cta((KG)^\circ) = 2 \cdot p^n + 2$.

<u>Case 2:</u> Let $char(K) = p$. In this case the group algebra is solvable as described within section 5.5.5. The maximal tori are exactly the radical complements and all are conjugated by the Wedderburn-Malcev theorem. Their dimension is $mto(KG) = 4$. The Cartan subalgebras are the centralizers of the radical complements and are all conjugate, too. Their dimension is $cta((KG)^\circ) = 2 \cdot p^n + 2$. An explicit construction is given within section 5.5.5 for one Cartan subalgebra. We use this construction to determine that all Cartan subalgebras are abelian: $ctn((KG)^\circ) = cts((KG)^\circ) = 1$. Let $a, b \in G$, $z := a^{p^n} = b^2$, $o(a) = 2p^n$ and $o(b) = 4$ such that $a^b = a^{-1}z$ and $G = \langle a, b \rangle$. If $H := \{1, b, b^2, b^3\}$, then KH is a maximal torus of KG and $C_{KG}(KH)$ is a Cartan subalgebra of $(KG)^\circ$. Within section 5.5.5 we have proven that $C_{KG}(KH)$ is linear spanned by the orbit-sums of the action of $H = \langle b \rangle$ on G. For every $i \in \mathbb{N}$ the elements a^i and $a^{-i}z$ as well as $a^i b$ and $a^{-i}bz$ are conjugates in G with respect to $b \in H$. We conclude that the elements of a basis of $C_{KG}(KH)$ are $1, z, z_j, z_j b, zz_j, zz_j b$ for several $j \in \mathbb{N}$ and central elements $z_j = a^j + a^{-j}z$. By using this result it is straightforward to prove that $C_{KG}(KH)$ is commutative.

<u>Case 3:</u> Let $char(K) = 2$. By using proposition 23 it is sufficient to analyze the dimension, the solvable and nilpotency class of the Cartan subalgebras for a finite field F_q of characteristic 2. N. Makhijani, R. K. Sharma and J. B. Srivastava have analyzed the group algebra of dihedral and quaternion groups for F_q in their articles [41] and [42]. For every divisor m of p^n there exist a natural number e_m (the exact definition of e_m is not needed in our context) such that the factor algebra $KG/rad(KG)$ is isomorphic to $A := F_q \oplus \bigoplus_{m | p^n, m > 1} ((F_{q^{e_m}})^{2 \times 2})^{\frac{\phi(m)}{2e_m}}$. As proven in example 12 we derive $mto(A) = mto(KG) = p^n$. A maximal torus of KG is $K\langle a^2 \rangle$. Thus we have to analyze the Cartan subalgebra $C_{KG}(\langle a^2 \rangle)$. For this the action of $\langle a^2 \rangle$ on G is to be determined. Every element of $\langle a \rangle$ is stable under the action of $\langle a^2 \rangle$. As proven for dihedral groups there are two G-conjugacy classes which are $\langle a^2 \rangle$-classes, too, and the centralizer is K-linear spanned by the maximal torus and these two classes. In particular, the centralizer is commutative and of dimension $p^n + 2$: for all fields K the dimension of the Cartan subalgebras of $(KG)^\circ$ is identical as well as the solvable and nilpotency class.◇

In general, the dimension of the Cartan subalgebras in modular Lie algebras is not unique. In the textbook of Seligman [58] there are some examples included on pages 116 and 117 such that there are Cartan subalgebras of dimension 1 and 2 as well of dimension $p^k - 1$ for several distinct natural numbers k.

We finish this section with some remarks on the weight decomposition. T. Bauer analyzed in his dissertation [4] the weights, the weight spaces and the weight decomposition for solvable associative algebras A possessing a factor algebra by the nilradical isomorphic to K^n for a natural number n. If $e_1, \cdots, e_n$ are the pairwise orthogonal idempotents of a radical complements of A, then A decomposes by the two-sided Pierce decomposition into the Pierce components $e_i A e_j$. The Cartan subalgebra (see theorem 26) is $\bigoplus_{i=1}^{n} e_i A e_i$. Bauer proves that the non-zero Pierce components $e_i A e_j$ with $i \neq j$ are the other weight spaces, and he provides explicit constructions of the corresponding weights. By using theorem 26 all Cartan subalgebras are conjugated. Thus the weight space decomposition is unique in this case as proven within lemma 14 for Lie algebras based on associative algebras over algebraically closed fields. An explicit construction of the weights and weight spaces is not known to the author for this special case. In general, it is not known whether a Lie algebra based on an associative algebra possesses a weight space decomposition over an arbitrary field. The author conjectures that such a decomposition always exists and that the dimensions of the weight spaces are independent of the choice of the Cartan subalgebra. By lemma 14 it is sufficient – for the proof of the independence – to analyze that a weight space decomposition is compatible with base field extension.

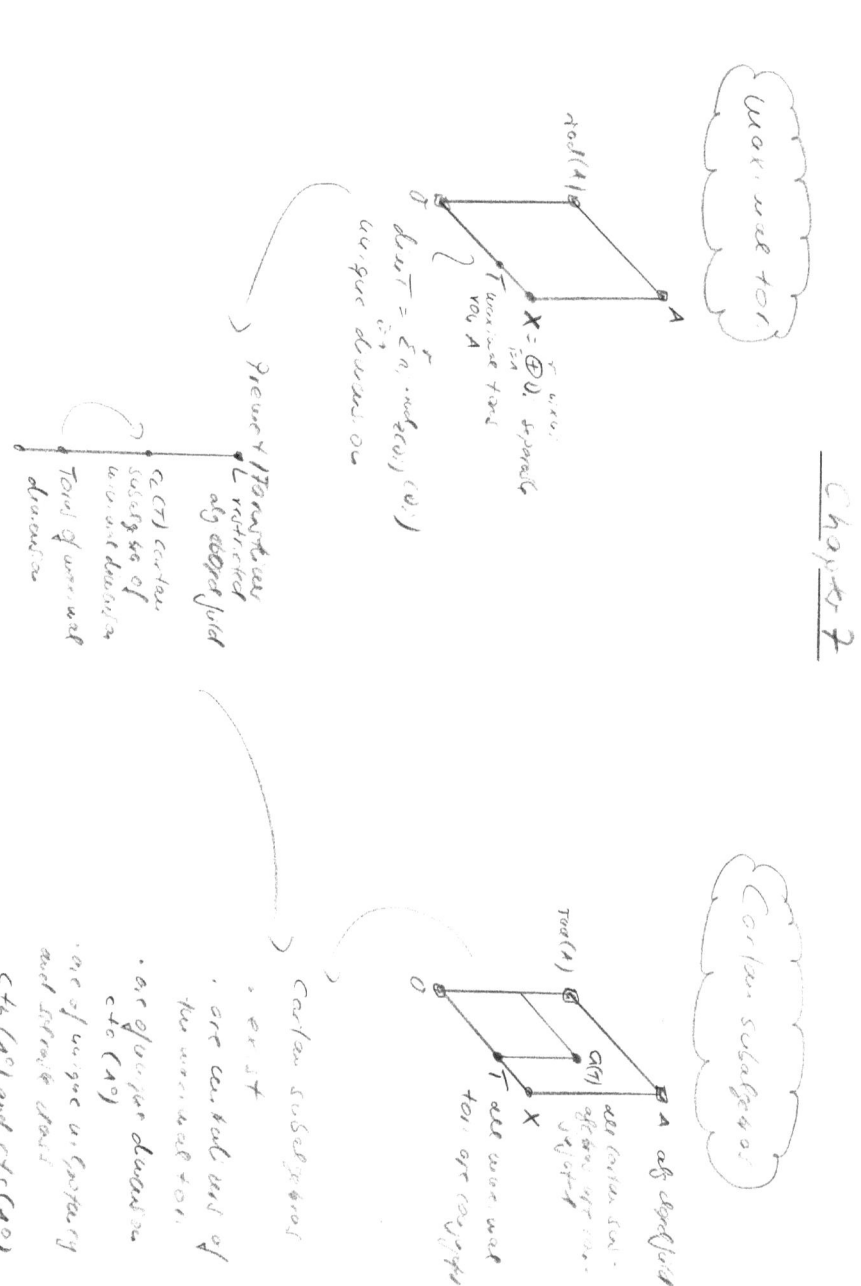

7.3 Open-ended questions and exercises

Open-ended questions 11 *(i) Is the dimension of the maximal tori unique without the assumption that the factor algebra by its nilradical is separable? Within [92] a proof is given for arbitrary finite-dimensional associative division algebras based on linear algebraic groups. For solving this open-ended question the case of an arbitrary matrix algebra over a finite-dimensional division algebra is to be analyzed.*

(ii) Determine mto for tensor products!

(iii) Determine mto for matrix algebras over algebras!

(iv) Determine mto for group algebras!

(v) Determine a formula for cta in associative algebras!

(vi) Determine ctd for group algebras!

(vii) Determine cts for group algebras!

(viii) Determine ctn for group algebras!

(ix) Determine all finite groups G such that $ctn((KG)^\circ)$ is unique for all choices of a field K. Examples are dihedral resp. quaternion groups of order $2p^n$ and $4p^n$ resp. $4p^n$ for $n \in \mathbb{N}$ and p being an uneven prime number.

(x) Determine the nilpotency and solvable class of the Lie algebra associated to a tensor product of two Lie nilpotent associative algebras! With respect to corollary 54 we conjecture that it is the maximum of the corresponding classes of the factors ($A^{n \times n}$ is isomorphic to $A \otimes K^{n \times n}$).

(xi) Determine a weight decomposition with respect to a Cartan subalgebra of the Lie algebra associated to a finite-dimensional associative unitary K-algebra. In addition, determine the dimension of the corresponding weight spaces. Which associative algebras possess a weight space decomposition?

(xii) Determine those finite groups G such that $mto(KG)$ is identical for every field K.

(xiii) Does a corresponding weight decomposition in the group of units exists with respect to a Carter subgroup? Is this decomposition somehow unique?

(xiv) Within lemma 13 analyze whether the dimension of the Cartan subalgebra is maximal for a maximal torus of minimal dimension.

Let A be a finite-dimensional associative unital algebra with separable factor algebra by its nilradical, $r, n_1, \ldots, n_r \in \mathbb{N}$ and $D_1, \ldots, D_r$ division algebras such that $A/rad(A)$ is isomorphic to $\bigoplus_{i=1}^{r} D_i{}^{n_i \times n_i}$. A is called homogenous of type I if $n_i = n$ is valid for all $i \in \underline{r}$. A is called homogenous of type II if $D_i = D$ is valid for all $i \in \underline{r}$. A is called homogenous of type III if $n_i = n$ and $D_i = D$ are valid for all $i \in \underline{r}$.

Excercise 334 *Prove proposition 22 in details.*

Excercise 335 *Determine mto(A) for A being local.*

Excercise 336 *Determine mto(A) for A being basic.*

Excercise 337 *Determine mto for direct products.*

Excercise 338 *Determine mto(A) for A being central ($Z(A) = K \cdot 1_A$).*

Excercise 339 *Determine mto(A) for A being a central division algebra.*

Excercise 340 *Determine mto(A) for A being a central-simple algebra.*

Excercise 341 *Determine mto(A) for A being a separable simple algebra.*

Excercise 342 *Determine mto(A) for A being a separable division algebra.*

Excercise 343 *Determine mto(A) for A being a central separable algebra.*

Excercise 344 *Determine mto(A) for A being a separable algebra.*

Excercise 345 *Determine mto(A) for A being homogenous of type I.*

Excercise 346 *Determine mto(A) for A being homogenous of type II.*

Excercise 347 *Determine mto(A) for A being homogenous of type III.*

Excercise 348 *Determine mto(A) for A being a semisimple group algebra.*

Excercise 349 *Determine mto(A) for A being a semisimple group algebra based on a dihedral group.*

Excercise 350 *Determine mto(A) for A being a semisimple group algebra based on a quaternion group.*

Excercise 351 *Determine mto(A) for A being a semisimple group algebra based on a semi-dihedral group.*

Excercise 352 *Determine $mto(A)$ for A being a quaternion algebra in characteristic 2.*

Excercise 353 *Determine $mto(A)$ for A being a quaternion algebra in characteristic $\neq 2$.*

Excercise 354 *Determine $mto(A)$ for A being a full matrix algebra over a field.*

Excercise 355 *Determine $mto(A)$ for A being a full matrix algebra over K.*

Excercise 356 *Determine $mto(A)$ for A being a Solomon algebra in characteristic zero.*

Excercise 357 *Determine $mto(A)$ for A being a Solomon-Tits algebra.*

Excercise 358 *Determine $mto(A)$ for A being the subalgebra of lower triangular matrices in $K^{n\times n}$.*

Excercise 359 *Determine $mto(A)$ for A being the subalgebra of upper triangular matrices in $K^{n\times n}$.*

Excercise 360 *Formulate an analogue to theorem 48 for non-unitary algebras and prove it!*

Excercise 361 *Prove the statement $C_A(T) = T \oplus C_{rad(A)}(T)$ mentioned before theorem 48!*

Excercise 362 *Prove that an algebraically closed field is infinite. (Tip: Focus on the polynomial $1 + \prod_{k \in K}(t-k)$ for a finite field K. Do you see a similarity to the proof that infinite many prime numbers exist?)*

Excercise 363 *Let K be a field, p a prime number, A a finite abelian group with $(\mid A \mid, p) = 1$ and P a finite p-group. Focus on the group $G := P \times A$. Find two choices of the field K such that the dimensions of the Cartan subalgebras of $(KG)^\circ$ is different for both choices.*

Excercise 364 *Let K be a field, p a prime number and P a finite p-group. Find two choices of the field K such that the dimensions of maximal tori of $(KG)^\circ$ is different for both choices.*

Excercise 365 *Is proposition 23 valid for an arbitrary associative algebra?*

Excercise 366 *Determine $cta(\cdot)$ for the Solomon-Tits algebra.*

Excercise 367 *Determine $cta(\cdot)$ for the Solomon algebra in characteristic zero.*

Excercise 368 *Determine $cta(\cdot)$ for the algebra of lower triangular matrices over a field.*

Excercise 369 *Determine $cta(\cdot)$ for the algebra of upper triangular matrices over a field.*

Excercise 370 *Determine $cta(\cdot)$ for a full matrix algebra over a field.*

Excercise 371 *The symmetric group S_3 is a Frobenius group.*

Excercise 372 *Prove example 10 in details!*

Excercise 373 *Analyze under what terms a solvable group algebra possesses a central nilradical and vice versa!*

Excercise 374 *True or false: An associative algebra possessing exactly one radical complement is solvable.*

Excercise 375 *True or false: An associative algebra possessing exactly one radical complement is Lie nilpotent.*

Excercise 376 *True or false: If an associative algebra possesses exactly one radical complement, then the radical complement is central.*

Excercise 377 *True or false: Let G be a finite group and K, L be fields. Then $mto(KG) = mto(LG)$ and $cta((KG)^\circ) = cta((LG)^\circ)$ are valid.*

Excercise 378 *Let K be a field and $G := A_4$. Prove that the dimension of maximal tori of KG and Cartan subalgebras of $(KG)^\circ$ is unique. In this exercise use the main theorems of this chapter.*

Excercise 379 *Let K be a field and $G := A_4$. Prove that the dimension of maximal tori of KG and Cartan subalgebras of $(KG)^\circ$ is unique. In this exercise prove the result as done in exercise 10.*

Excercise 380 *Let K be a field and $G := A_4$. By using [41] and its references determine $mto(KG)$ and $cta((KG)^\circ)$.*

Excercise 381 *Let K be a field and $G := S_4$. By using [41] and its references determine $mto(KG)$. (Tip: Analyze the cases of characteristic 3 (a radical complement is isomorphic to $K^{3\times 3} \times K^{3\times 3} \times K^2$), characteristic 2 (a non-commutative radical complement of dimension 5 exists) and the semisimple case.)*

Excercise 382 Let K be a field and $G := D_{12}$. By using [41] and its references determine $mto(KG)$ and $cta((KG)^\circ)$. (Tip: Analyze the cases of characteristic 3 (solvable case), characteristic 2 (central radical of dimension 7, non-commutative complement of dimension 5) and the semisimple case.)

Excercise 383 Let G, H be finite groups such that $cta((KG)^\circ$ and $cta((KH)^\circ$ is unique for all choices of a field K. Prove that $cta((K(G \times H))^\circ$ is unique for all choices of a field K, too.

Excercise 384 Let $n, r \in \mathbb{N}$, $p, q \in \mathbb{P}$, G be a finite group and K a field. Prove or disprove that $cta((KG))^\circ$ is unique for all choices of a field K:

(i) G is abelian.

(ii) G is a p-group.

(iii) G possesses a normal p-Sylow subgroup.

(iv) $G = D_{2p^n}$

(v) $G = Q_{4p^n}$

(vi) $G = D_{2p^n} \times Q_{4p^n}$

(vii) $G = D_{2p^n} \times Q_{4q^r}$.

Excercise 385 Prove that the nilpotency and solvable class of the Lie algebra associated to a tensor product of two Lie nilpotent associative algebras is not smaller than the maximum of the corresponding classes of each factor.

Excercise 386 Let K be a field and A an associative finite-dimensional unitary K-algebra with separable factor algebra by its nilradical. Prove that $mto(A) \leq cta(A^\circ)$ is valid. On what terms is the statement $mto(A) = cta(A^\circ)$ true? If the factor algebra is not separable, then prove that the maximal dimension of all maximal tori is not greater than $cta(A^\circ)$.

Excercise 387 Let K be a field, A an associative finite-dimensional unitary K-algebra and S a subalgebra of A. Find examples of K, A, s such that $cta(S^\circ) < cta(A^\circ)$, $cta(S^\circ) > cta(A^\circ)$ and $cta(S^\circ) = cta(A^\circ)$ is valid. Find analogue examples if S is a left ideal, a right ideal or an ideal of A.

Excercise 388 Determine cta, ctn and cts for the zero-extension (see exercise 39). If the analysis is to complex, assume e.g. that A is separable.

Excercise 389 Determine cta, ctn and cts for the algebra eAe (see exercise 136). If the analysis is to complex, assume e.g. that A is separable or that that e is central.

Excercise 390 Let p be a prime number, $n \in \mathbb{N}$, K a field and $G \in \{D_{2p^n}, D_{4p^n}, Q_{4p^n}\}$. Is the dimension, nilpotency and solvable class of the Cartan subalgebras of $(KG)^\circ$ the same as the corresponding ones for the nilradical of $(KG)^\circ$. (Tip: Focus on Lie products with generators of G)

Excercise 391 Let K be a finite field and G the $ax + b$-group with respect to K. For an arbitrary field L determine the Cartan subalgebras of $(LG)^\circ$. Is it possible to calculate their dimension, nilpotency and solvable class?

Excercise 392 Let p, q be prime numbers, G a pq-group and L a field. Determine the Cartan subalgebras of $(LG)^\circ$. Is it possible to calculate their dimension, nilpotency and solvable class?

Excercise 393 Is it possible to extend the results of exercises 391 and 392 to special Frobenius groups?

Excercise 394 Let K be a field and $G \in \{A_4, S_4\}$. Determine the Cartan subalgebras of $(LG)^\circ$. Is it possible to calculate their dimension, nilpotency and solvable class?

Excercise 395 Let K be a field. Determine $cta((KG)^\circ)$, $cts((KG)^\circ)$, $ctn((KG)^\circ)$ and $mto(KG)$ for $G := D_{2 \cdot 3 \cdot 5}$.

Excercise 396 Let K be a field. Determine $cta((KG)^\circ)$, $cts((KG)^\circ)$, $ctn((KG)^\circ)$ and $mto(KG)$ for $G := D_{2 \cdot 2 \cdot 3 \cdot 5}$.

Excercise 397 Generalize example 12 for the semisimple case and for the case of characteristic 2.

Excercise 398 Generalize example 13 for the semisimple case and for the case of characteristic 2.

Excercise 399 Generalize example 14 for the semisimple case and for the case of characteristic 2.

Excercise 400 Prove the add-on of lemma 14 in details!

Chapter 8

Outlook on series II

In volume II of this series we will restrict our analysis to solvable associative algebras. We have already described the Cartan subalgebras and the nilradical of the associated Lie algebra in this volume. These are two examples of special maximal nilpotent substructures. Therefor we want to describe and construct all maximal nilpotent Lie subalgebras. The nilradical and the Cartan subalgebras will crystallize to be extremal in special manner which we will point out in volume II.

In addition, we will study maximal nilpotent subgroups of the group of units. Both extremal subalgebras have their counterpart in the Fitting subgroup and the Carter subgroups. Also all other maximal nilpotent subgroups have a maximal nilpotent pendant in the Lie algebra. Again, the Fitting subgroup and the Carter subgroups are extremal in the same manner within all maximal nilpotent subgroups.

The correspondence of maximal nilpotent Lie subalgebras and subgroups – which can be formulated by the 1:1 bijection 'form group of units' and its inverse 'generate the linear span' – and the structure, characteristics and construction of the related maximal substructures are the main content of volume II. One of the important utilities for their study are single and double centralizers. The remarkable theorem of Du plays an significant rule for describing the connection of the class of nilpotency of a maximal nilpotent subalgebra and its counterpart in the group of units. Therefor we will introduce this theory in chapter one using an approach of H. Laue in [39].

Our results will be illustrated by our main examples: upper and lower triangular matrix algebras, Solomons algebras, Solomon-Tits algebras and solvable group algebras.

> 'That which is provable, ought not to be believed in science without proof.' (Richard Dedekind, 1888)

Outlook on Series 2

Focus: first-class extremal isomorphism extrema with its its algebra syntactic algebra and characteristic not equal to 2.

Maximal unipotent subgroups — extremal?

Maximal unipotent U_i's

Cartan subgroups — extremal?

Cartan subgroups Q_i's → normal

Filling subgroup — extremal?

Appendix A

Derived algebras

In this appendix we study a class of solvable associative algebras which are Lie nilpotent. In particular, we examine the conditions for two algebras being isomorphic within this class. Armin Jöllenbeck has studied these algebras within his diploma thesis [32]. One important aspect of his thesis was to analyze free algebras of this special class.

A.1 Definition and initial properties

Definition 3 *Let A be a K-algebra and $c, d \in K$. We define a multiplication $\circ_{c,d}$ on A by $a \circ_{c,d} b := cab + dba$ for all $a, b \in A$.*⋄

The following proposition is to be proved by the reader as an exercise (see exercise 422):

Proposition 24 *Let A, B be K-algebras and $c, d, e, f \in K$. The following statements are valid:*

(i) $(A; +; \circ_{c,d})$ *is a K-algebra, indicated by $A_{c,d}$ and called the derived algebra of A by $\circ_{c,d}$.*

(ii) $A = A_{1,0}$ *(A itself)*

(iii) $A^- = A_{0,1}$ *(the inverse algebra of A)*

(iv) $A^\circ = A_{1,-1}$ *(the Lie algebra associated to A)*

(v) $A_{1,1}$ *(the Jordan-algebra associated to A)*

(vi) $A_{0,0}$ *is a zero-algebra (The multiplication of two elements is always zero).*

(vii) *If A, B are isomorphic, then $A_{c,d}, B_{c,d}$ are isomorphic, too.*

(viii) *Every homomorphism of A is a homomorphism of $A_{c,d}$, too.*⋄

Proposition 25 *Let A be an associative K-algebra and $c, d \in K$. If the condition $A^2 \subseteq Z(A)$ is assumed, then $(A_{c,d})^2 \subseteq Z(A_{c,d})$ is valid, too, and $A_{c,d}$ is associative. In particular, A and $A_{c,d}$ are solvable algebras.*

Proof. The statement $(A_{c,d})^2 \subseteq Z(A_{c,d})$ and the associativity of $A_{c,d}$ is to be proven by the reader as an exercise (see exercise 426). The condition $A^2 \subseteq Z(A)$ for an algebra A implies that $Z(A)$ is an ideal of A and that $A/Z(A)$ is a zero-algebra. Using the commutativity of $Z(A)$ we deduce that A is solvable. ⋄

Example 15 *The set of strict upper triangular matrices with respect to a field K in $K^{3\times 3}$ satisfy the preconditions of proposition 25.*⋄

Proposition 26 *Let A, B be K-algebras and $c, d, e, f \in K$. The following statements are valid:*

(i) $(A_{c,d})_{e,f} = A_{c+e,d+f}$ *(iterative derivation of A by $\circ_{c,d}$ and $\circ_{e,f}$)*

(ii) *The operation δ on K^2 defined by $(c,d)\delta(e,f) := (c+e, d+f)$ results into a monoid on K^2 with unity element $(1,0)$.*

(iii) *(c,d) is invertible with respect to δ if and only if $c^2 - d^2 \neq 0$ is valid.*

(iv) *Let (c,d) be invertible with respect to δ. Then the algebras $A_{c,d}$ and $B_{c,d}$ are isomorphic if and only if A and B are isomorphic.*

Proof. The detailed proof of this proposition is left to the reader. Parts (i)-(iii) are to be proved by direct calculations. One part of (iv) is a consequence of part (vii) of proposition 24. Now let (c,d) be invertible with inverse (e,f). If $A_{c,d}$ and $B_{c,d}$ are isomorphic, then the derived algebras by (e,f) are isomorphic, too. These are exactly A and B because (e,f) is the inverse of (c,d). Therefore $A = A_{1,0}$ and $B = B_{1,0}$ are valid using part (ii).⋄

Proposition 27 *Let A be a K-algebra and $c, d, e, f \in K$. The following statements are valid:*

(i) *If $e \neq 0$ is assumed, then $A_{e,f}$ and $A_{1,\frac{f}{e}}$ are isomorphic.*

(ii) *If $f \neq 0$ is assumed, then $A_{e,f}$ and $A_{\frac{e}{f},1}$ are isomorphic.*

(iii) *If $c \neq 0$ is assumed, then $A_{ce,cf}$ and $A_{e,f}$ are isomorphic.*

(iv) *If $k \neq 0$ is assumed, then $A_{k,1}$ and $A_{1,\frac{1}{k}}$ are isomorphic.*

Proof. The detailed proof of this proposition is left to the reader. The parts (i)-(iii) can be proven by using the multiplication with e, f and c. On the one hand – using part (ii) and the fraction with k – $A_{k^2,k}$ is isomorphic to $A_{k,1}$. On the other hand using part (i) we deduce the isomorphism of these two algebras to $A_{1,\frac{1}{k}}$, and we have proven part (iv).⋄

A.2 Structural properties

Proposition 28 *Let A be an associative K-algebra, $c, d \in K$, $a \in A$ and $n \in \mathbb{N}_{\geq 2}$. a to the power of n with respect to $\circ_{c,d}$ is exactly $(c+d)^{n-1}a^n$.*

Proof. A first calculation implied $a \circ_{c,d} a = ca^2 + da^2 = (c+d)a^2$. Now the proof is to be finished by induction with respect to n. ⋄

Theorem 54 *Let A be a finite-dimensional associative K-algebra over a field K, $c, d \in K$ and $A^2 \subseteq Z(A)$. The following statements are true:*

(i) $rad(A) = Nil(A) \subseteq rad(A_{c,d}) = Nil(A_{c,d})$

(ii) In the case of $c \neq -d$ we deduce $rad(A) = rad(A_{c,d})$

(iii) In the case of $c = -d = 0$ the algebra $A_{c,d}$ is a zero-algebra. Every derived algebra of a zero-algebra is a zero-algebra, too.

(iv) Let $c = -d \neq 0$ and A be commutative. Then $A_{c,d}$ is a zero-algebra.

(v) Let $c = -d \neq 0$ and A be non-commutative. $A_{c,d}$ is nilpotent and all Lie triple products are zero.

Proof. By proposition 25 all indicated algebras are solvable. Using proposition 15 we deduce that $rad(A)$ is exactly the set of nilpotents elements of A. Now we can finish the proof by straightforward calculations and by using proposition 28. The detailed proof of this proposition is again left to the reader (see exercise 428). ⋄

A.3 Lie nilpotency

Proposition 29 *Let A be an associative K-algebra over a field K, $c, d \in K$ and $A^2 \subseteq Z(A)$. The algebras A° and $(A_{c,d})^\circ$ as well as the quasi-regular group of A and $A_{c,d}$ are nilpotent. Their class of nilpotency is at most 2.*

Proof. A^2 is central in A, and thus $A \circ A$ is central in A°, too. We deduce that $(A \circ A) \circ A = 0$ is true, and therefor the class of nilpotency of A is at most 2. A similar argument is valid for the quasi-regular groups: commutators of quasi-regular elements are elements of A^4, and therefor the commutator subgroup is central in $A^\star$. We conclude that $cl(A^\star) \leq 2$ is true. The condition that squares are central is passed from A to $A_{c,d}$. Now we can apply the results already proven for A on $A_{c,d}$. ⋄

A.4 Isomorphism of derived algebras

Initial position

Let A be an associative K-algebra over a field K, $c, d, e, f \in K$ and we assume $A^2 \subseteq Z(A)$. Using theorem 54 we have to analyze the following cases for studying the question of isomorphism between $A_{c,d}$ and another derived algebra $A_{e,f}$ of A:

$c \neq -d$ (1.1)
$c = -d = 0$ $A_{c,d}$ zero-algebra (1.2.1)
$c = -d \neq 0$ A commutative, $A_{c,d}$ zero-algebra (1.2.2)
$c = -d \neq 0$ Lie nilpotent of nilpotency class 2, A non-commutative (1.3)

$e \neq -f$ (2.1)
$e = -f = 0$ $A_{e,f}$ zero-algebra (2.2.1)
$e = -f \neq 0$ A commutative, $A_{e,f}$ zero-algebra (2.2.2)
$e = -f \neq 0$ Lie nilpotent of nilpotency class 2, A non-commutative (2.3).

We study the problem of isomorphism for all possible combinations between $A_{c,d}$ and $A_{e,f}$.◇

1.2 versus 2.2

If the preconditions $c = -d \neq 0$ and $e = -f \neq 0$ are valid, then the corresponding derived algebras are isomorphic:
$\varphi : x \mapsto \frac{c}{e}x$ is a K-linear isomorphism between $A_{c,d}$ and $A_{e,f}$. Obviously, the function is K-linear. For all $x, y \in A$ we calculate: $(x \circ_{c,d} y)\varphi = (cxy + dyx)\varphi = \frac{c^2}{e}xy + \frac{cd}{e}yx$ and $(x\varphi) \circ_{e,f} (y\varphi) = \frac{c^2}{e}xy + \frac{fc^2}{e^2}yx$. The identity $\frac{cd}{e} = \frac{fc^2}{e^2}$ is valid because its equivalent to $d = \frac{fc}{e}$ and to $\frac{d}{c} = \frac{f}{e}$. These statements are -1 using our assumption. Thus there is only one class of isomorphism in this case: $A_{1,-1}$ is the associated Lie algebra in which all squares are zero.◇

1.2 versus 2.1

Under the assumptions $c = -d \neq 0$ and $e = -f = 0$ both algebras are isomorphic if and only if A is commutative:
$A_{e,f}$ is a zero-algebra. Furthermore the identity $x \circ_{c,d} y = c(xy - yx)$ is valid for all $x, y \in A$. Both algebras are isomorphic if and only if $A_{c,d}$ is a zero-algebra, too. The condition $c \neq 0$ implies that this is only the case for A being commutative.◇

1.2 versus 2.3

We consider the cases $A_{1,-1}$ (see A.4) and $A_{e,f}$ with $e + f \neq 0$.

Case 1: $A_{1,-1}$ is a zero-algebra.
Both algebras are isomorphic if and only if $A_{e,f}$ is a zero-algebra, too. We prove that this is the case only for A being a zero-algebra or $e = f$ and $x^2 = 0$ are valid for all $x \in A$.

We suppose that A is a zero-algebra. This implies that $A_{e,f}$ is a zero-algebra, too, and both zero-algebras are isomorphic as they have the same dimension.
Now let $x^2 = 0$ for all $x \in A$ and $e = f$. Using this assumption on $(x+y)^2 = 0$ we conclude $xy + yx = 0$ for all $x, y \in A$. Therefore the multiplication in $A_{e,f}$ is given by $(e - f)xy$ for all $x, y \in A$. The condition $e = f$ implies that $A_{e,f}$ is indeed a zero-algebra.

Now let $A_{e,f}$ be a zero-algebra. Then the identity $0 = x \circ_{e,f} x = (e + f)x^2$ is valid for all $x, y \in A$. By using the condition $e + f \neq 0$ we conclude that all squares are zero in A. By $0 = (x + y)^2$ the identity $xy + yx = 0$ is valid for all $x, y \in A$. Therefor the multiplication in $A_{e,f}$ can be calculated by $(e - f)xy$ for all $x, y \in A$. Thus $A_{e,f}$ is a zero-algebra if and only if $e = f$ is valid or A is a zero-algebra.

Case 2: $A_{1,-1}$ is no zero-algebra.
We deduce that both algebras are isomorph if and only if $x^2 = 0$ for all $x \in A$ and $e - f \neq 0 \neq 2$ are valid.

Let both algebras be isomorphic. Squares are zero in $A_{1,-1}$, hence they are zero in $A_{e,f}$, too. This implies $0 = (e + f)x^2$ for all $x \in A$. By using the assumption $e + f \neq 0$ we conclude $x^2 = 0$ and therefor $xy + yx = 0$ for all $x, y \in A$. Thus the multiplication in $A_{e,f}$ is to be calculated by $(e - f)xy$ for all $x, y \in A$. By our assumption $A_{e,f}$ is no zero-algebra, and therefor the condition $(e - f) \neq 0$ is valid. $A_{1,-1}$ is not a zero-algebra in this case. Using $xy + yx = 0$ for all $x, y \in A$ we get for its multiplication $2xy$ for all $x, y \in A$. This implies $2 \neq 0$.

Now let $x^2 = 0$ for all $x \in A$ and $e - f \neq 0 \neq 2$. The multiplication in $A_{1,1}$ resp. in $A_{e,f}$ is represented by $2xy$ resp. $(e - f)xy$ for all $x, y \in A$. The multiplication with $\frac{2}{e-f}$ is an isomorphism from $A_{1,1}$ onto $A_{e,f}$. ◇

1.1 versus 2.1

In this case both algebras are isomorphic because they are zero-algebras of the same dimension.⋄

1.1 versus 2.3

The problem of being isomorphic is reduced to the question for $A_{e,f}$, $e+f \neq 0$ being a zero-algebra. This problem is already analyzed.⋄

1.1 versus 2.2

The problem of being isomorphic is reduced to the question for $A_{e,f}$, $e = -f \neq 0$ being a zero-algebra. The multiplication in $A_{e,f}$ is to be calculated by $e(xy - yx)$ for all $x, y \in A$. Hence using the fact $e \neq 0$ we deduce that $A_{e,f}$ is a zero-algebra if and only if A is commutative.⋄

1.3 versus 2.1

This problem is already analyzed.⋄

1.3 versus 2.2

This problem is already analyzed.⋄

1.3 versus 2.3

The case of commutativity:

We begin studying the question for $A_{e,f}$ being commutative. This is equivalent to the condition $(e - f)(xy - yx) = 0$ for all $x, y \in A$. Therefor $A_{e,f}$ is commutative if and only if $e = f$ is valid or A is commutative.
$A_{c,c}$ is isomorphic to $A_{1,1}$ by the multiplication with c for $c \neq 0$. In the case $c = 0$ we are facing again the case of zero-algebras which we have already analyzed.

If $A_{c,c}$ is isomorphic to $A_{e,f}$, then $A_{e,f}$ is commutative. Hence $e = f$ is valid or A is commutative. In the case $e = f \neq 0$ both algebras are isomorphic to $A_{1,1}$.
Now let A be commutative. Then the multiplication of $A_{1,1}$ is exactly $2xy$ and the one of $A_{e,f}$ can be calculated by $(e+f)xy$ for all $x, y \in A$. As both algebras are not zero-algebras we can assume $2 \neq 0 \neq (e+f)$. In this case the multiplication with $\frac{e+f}{2}$ is an isomorphism between both algebras.

We have to study the following question:
On what terms is $A_{c,d}$ isomorphic to $A_{e,f}$ for $c + d \neq 0 \neq c - d$ and

$e - f \neq 0 \neq e + f$?

We want to reduce this problem to the algebras $A_{1,0}$, $A_{0,1}$ and $A_{1,k}$ for a $k \neq 0$. Proposition 27 lets us derive that $A_{e,f}$ for $e \neq 0$ is isomorphic to $A_{1,f/e}$, $A_{e,f}$ for $f \neq 0$ to $A_{e/f,1}$ and $A_{k,1}$ for $k \neq 0$ to $A_{1,\frac{1}{k}}$. $e + f \neq 0$ is invariant under an isomorphism between $A_{1,f/e}$ and $A_{e/f,1}$. Hence we have to study the algebras $A_{1,0}$, $A_{0,1}$ and $A_{1,k}$ for $k \neq 0, 1, -1$ (1 was already analyzed within the case of commutativity!).

We are reducing the question of isomorphism further.
We consider the case of $A_{0,1}$ and $A_{1,k}$ with $k \neq 0, 1, -1$.
$(0, 1)$ is invertible with respect to δ (its an involution). Using proposition 26 we conclude that the problem is equivalent to the isomorphism-problem between $A_{1,0}$ and $A_{k,1}$. Proposition 27 implies that $A_{k,1}$ and $A_{1,k^{-1}}$ are isomorphic. Therefor we have transferred the problem to the isomorphism-problem between $A_{1,0}$ and $A_{1,k}$ with $k \neq 0, 1, -1$.

Now we consider the case $A_{1,k}$ and $A_{1,l}$ with $k \neq 0, 1, -1$ and $l \neq 0, 1, -1$
We derive both algebras by $(1, -k^{-1})$. Because of $1 - k^{-2} \neq 0$ the isomorphism-problem is unaltered using proposition 27. The derivation of $A_{1,k}$ is $A_{0,k+k^{-1}}$ with $k + k^{-1} \neq 0$. $A_{1,l}$ is derived to $A_{1-lk^{-1},l-k^{-1}}$, and both components are non-zero.
Using the isomorphism of proposition 27 the first algebra is isomorphic to $A_{0,1}$ isomorph and the second one to $A_{1,x}$ with $x \neq 0$. Hence we have reduced the problem further.

The following isomorphism-questions remain: $A_{1,0}$ to $A_{1,k}$ and $A_{1,0}$ to $A_{0,1}$.

The case $A_{1,0}$ to $A_{0,1}$ cannot be solved here. Its the question under what terms an algebra is isomorphic to its inverse-algebra. In the literature there are prominent algebras satisfying this condition (e.g. quaternion-algebras, group algebras, matrix algebras). But there are also well-known examples not fulfilling this condition (e.g. Solomons algebras, Solomon-Tits algebras).

The case $A_{1,0}$ to $A_{1,k}$ can be solved using the theory of Armin Jöllenbeck related to free algebras (see [32]). Both algebras are not isomorphic if the algebras are associative. Otherwise $(1, 0)$ would be a multiple of $(1, k)$ which is impossible.◇

Appendix

Acid

- C→d — 1.1
- C=d=o — 1.2
- C=d≠o — 1.3

Isomorphism?

- 2.1 — e≠d
- 2.2 — e=d
- 2.3 — e≠d≠o

- 1.1 → commutator
- 1.2 → commutator / non-commutator
- 1.3 → non-commutator

A.5 Open-ended questions and exercises

Open-ended questions 12 *(i) Find necessary and sufficient conditions for an algebra A being isomorphic to its inverse-algebra A^-?*

(ii) Solve the first question under the precondition $A^2 \leq Z(A)$!

(iii) Solve the isomorphic-question – $A_{1,0}$ vs. $A_{1,k}$ – without using the associativity of A!

Excercise 401 *Within this chapter there are several isomorphism given related to the multiplication with an element. Prove that all these functions are isomorphism!*

Excercise 402 *True or false: Every algebra is isomorphic to its inverse-algebra. Specify well-known examples for this question!*

Excercise 403 *Let A be a K-algebra and $c, d \in K$. Are $A_{c,d}$ and $A_{d,c}$ isomorphic?*

Excercise 404 *Let A be a K-algebra and $c, d \in K$. True or false: $A_{d,c}$ is the inverse-algebra of $A_{c,d}$.*

Excercise 405 *Let A be a K-algebra and $c, d \in K \setminus \{0\}$. True or false: $A_{1/c,1/d}$ is the inverse-algebra of $A_{c,d}$.*

Excercise 406 *Let A be a K-algebra and $c, d \in K \setminus \{0\}$. True or false: $A_{1/c,1/d}$ is isomorphic to $A_{c,d}$.*

Excercise 407 *Prove proposition 26!*

Excercise 408 *Let K be a field. Define a monomorphism of $(K^2; \delta)$ into $K^{2\times 2}$!*

Excercise 409 *Let K be a field. Calculate the inverse of an invertible element of $(K^2; \delta)$ (Tip: previous exercise)!*

Excercise 410 *Let K be a field of characteristic zero. True or false:*

  (i) $A_{2,3}$ is isomorphic to $A_{1,3/2}$.

  (ii) $A_{3,2}$ is isomorphic to $A_{3/2,1}$.

  (iii) $A_{12,8}$ is isomorphic to $A_{3,2}$.

  (iv) $A_{7,1}$ is isomorphic to $A_{1,1/7}$.

  (v) $A_{1,-1}$ is isomorphic to $A_{5,-5}$.

(vi) $A_{7,-7}$ is isomorphic to $A_{5,-5}$.

(vii) $A_{1,-1}$ is isomorphic to $A_{0,0}$. What is the solution for $char(K) = 2$?

(viii) $A_{1,1}$ is isomorphic to $A_{0,0}$.

(ix) $A_{1,1}$ is isomorphic to $A_{3,3}$.

Are the elements $(2,3)$ etc. invertible with respect to δ? For the invertible pairs calculate their inverse!

Excercise 411 *Let K be a field of characteristic zero. Reduce the following problems of being isomorphic (as done in section A.4): $A_{0,1}$ vs. $A_{1,7}$ and $A_{1,6}$ vs. $A_{1,11}$.*

Excercise 412 *Let A be a K-algebra, $c,d \in K$ and $n \in \mathbb{N}$. Describe the n-fold derivation of A with (c,d)? (Tip: Use the representation of 2×2 - matrices).*

Excercise 413 *Let K be a field and A be an algebra. True or false: There is a unit (c,d) in $(K^2; \delta)$ such that A is isomorphic to $A_{c,d}$.*

Excercise 414 *Let K be a field. Is every element of $(K^2; \delta)$ invertible? Does a field exist for which this statement is true?*

Excercise 415 *Let K be a field. Find all involutions in $(K^2; \delta)$!*

Excercise 416 *Prove proposition 27!*

Excercise 417 *Let A be a K-algebra and $c,d \in K$. Find equivalent conditions for $A_{c,d}$ being commutative!*

Excercise 418 *Let A be a K-algebra and $c,d \in K$. Find equivalent conditions for $A_{c,d}$ being associative!*

Excercise 419 *Let A be a K-algebra and $c,d \in K$. Find equivalent conditions for $A_{c,d}$ being unitary!*

Excercise 420 *Let A be a K-algebra and $c,d \in K$. Find equivalent conditions for $A_{c,d}$ being a zero-algebra!*

Excercise 421 *Let A ba a K-algebra and $c,d \in K$. If A is a zero-algebra, then $A_{c,d}$ is a zero-algebra, too!*

Excercise 422 *Prove all statements within proposition 24 which are dedicated to be proven by the reader!*

Excercise 423 Let A, B be K-algebras and $c, d \in K$. True or false: Every homomorphism of $A_{c,d}$ into B is also one of A into B?

Excercise 424 Let A, B be K-algebras and $c, d \in K$. True or false: If $A_{c,d}$ and $B_{c,d}$ are isomorphic, then A and B are isomorphic, too. Is the statement true if (c, d) is invertible?

Excercise 425 Prove example 15! Find some more examples which satisfy the preconditions of proposition 25!

Excercise 426 Prove proposition 25 in details!

Excercise 427 Prove proposition 28!

Excercise 428 Prove theorem 54!

List of Figures

graphic for chapters 2 and 3	31
graphic for section 4.1	45
graphic for sections 4.3 and 4.4	51
graphic for sections 4.5-4.7	59
graphic for section 5.1	71
graphic for section 5.2	79
graphic for section 5.3	85
graphic for section 5.4	90
graphic for section 5.5, part 1	96
graphic for section 5.5, part 2	105
graphic for section 5.6	114
graphic for section 5.7	122
graphic for section 5.8	137
graphic for section 5.9	147
graphic for section 5.10	154
graphic for chapter 6	178
graphic for chapter 7	206
graphic for chapter 8, outlook on volume II	215
graphic for the appendix	223

Bibliography

[1] S. A. Amitsur: Finite subgroups of division rings, Trans. Amer. Math. Soc. 80(1955), pp. 361-386

[2] D.W. Barnes, On Cartan subalgebras of Lie algebras, Math. Z. 101, 1967, pp. 350-355

[3] B. G. Basmaji, Complex represenatations of metycyclic groups, American math. monthly, vol. 86, 1, 1979, pp. 47-48

[4] T. Bauer, Über die Struktur der Solomon-Algebren, Bayreuther Mathematische Schriften, Heft 63, 2001, 1-102

[5] T.Bauer, S. Siciliano, Carter subgroups in the group of units of an associative algebra, Bulletin of the Australian Mathematical Society, Volume 71, Issue 03, June 2005, pp. 471-478

[6] Dieter Blessenohl, Manfred Schocker, Noncommutative Character Theory of the Symmetric Group, Imperial College Press, United Kingdom, 2005

[7] G. Benkart, Cartan subalgebras in Lie algebras of Cartan type, Canadian Mathematical Society Conference Proceedings 5, 1986, pp.157-187

[8] Yakov Berkovitch, Avinoam Mann, On sums of degrees of irreducible characters, Journal of Algebra, 199, 1998, pp. 646-665

[9] Yaroslav Bezverkhny, Sum of cubes of degrees of irreducible complex characters, Ottawa-Carleton institute of mathematics and statistics, April 2003, Master of science

[10] N. Bourbaki, Groupes et algèbres de Lie, Chapitre VII, Hermann, Paris, 1973

[11] A.A. Bovdi, I.I. Khripta, Generalized Lie nilpotent group rings, Math. USSR Sbornik 57(1), 165-169, 1987

[12] Carter, R., On a class of finite soluble groups, Proc. London Math. Soc. 9, 1959, pp. 623-640

[13] R. Dedekind, Über Gruppen, deren sämmtliche Theiler Normaltheiler sind, Mathematische Annalen 1897, Band 48.4, Seiten 548-561

[14] Di Martino, L. - Tamburini, M.C. - Zalesskiĭ, A.E., Carter subgroups in classical groups, J. London Math. Soc. 55, 1997, pp. 264-276

[15] Rolf Farnsteiner, Varieties of tori and Cartan subalgebras of restricted Lie algebras, Trans. Amer. Math. Soc. 356, 2004, 4181-4236

[16] Gutan, Marin - Kisielewicz, Andrezej, Reversible group rings, Journal of algebra 279, 2004, pp. 280-291

[17] Gorenstein, D., Finite Groups, Harper and Row, New York, 1968

[18] R. Heffernan and D. MacHale, On the sum of the character degrees of a finite group, Department of Mathematics, University College Cork, Ireland, Mathematical Proceedings of the Royal Irish Academy, 2008

[19] Hallahan, C.H. - Overbeck, J.: Cartan subalgebras of meta-nilpotent Lie algebras, Math. Z. 116, 1970, pp. 215-217

[20] I.N. Herstein, Finite Multiplicative Subgroups In Division Rings, Pacific J. Math. 3, 1953, pp. 121- 126

[21] I.N. Herstein, Topics in Ring Theory, University of Chicago Press, Chicago, 1969

[22] I.N. Herstein, Lie and Jordan structures in simple associative rings, Bull. AMS, Vol. 67, No. 6, 1961, pp. 517-531

[23] I.N. Herstein, On the Lie structure of an associative ring, Journal of algebra, Vol. 14, Issue 4, April 1970, pp. 561-571

[24] L.K. Hua, Some properties of s-fields, Proc. Nat. Acad. Sci. U.S.A. 35, 1949, pp. 533-537

[25] B. Huppert, Endliche Gruppen I, Springer-Verlag, Berlin, 1967

[26] B. Huppert, N. Blackburn, Finite groups II, Springer-Verlag, Berlin, Heidelberg, New York, 1982

[27] I.M. Isaacs, Algebra, a graduate course, Brooks/Cole Publishing Company, Pacific Grove, California, 1993

[28] G. Ivanyos, Finding the radical of matrix algebras using Fitting decomposition, J. Pure Appl. Algebra 139, 1999, pp. 159-182

[29] N. Jacobson, Lie algebras, Wiley Interscience, New York London, 1962

[30] N. Jacobson, Schur's theorems on commutative matrices, Bull. Amer. Math. Soc. Volume 50, Number 6, 1944, pp. 431-436

[31] S.A. Jennings, Central chains of ideals in an associative ring, Duke Math.J 9, 1942, pp. 341-355

[32] Armin Jöllenbeck, Abgeleitete Algebren, Mathematisches Seminar der Christian-Albrechts-Universität zu Kiel, Diplomarbeit, 1994

[33] G. Karpilovsky, The Jacobson Radical Of Group Algebras, Elsevier, Amsterdam, 1987

[34] Gregory Karpilovsky, Unit groups of classical rings, Clarendon Press, Oxford, 1988

[35] Adalbert Kerber, Zu einer Arbeit von J. L. Berggren über ambivalente Gruppen, Pacific Journal of Mathematics, Vol. 33, No. 3, 1970

[36] I.I. Khripta, The nilpotence of the multiplicative group ring, Mat. Zametki 11, 1972, pp. 191-200

[37] Max-Albert Knus u.a., The book of involutions, AMS Colloquium Publications, Volume 44, 1998

[38] T. Y. Lam, Finite Groups Embeddable in Division Rings, www.arxiv.org

[39] H. Laue, Assoziative Algebren, Vorlesung am Mathematischen Seminar der CAU zu Kiel, WS 2010/2011

[40] H. Laue, Lie-Algebren, Vorlesung am Mathematischen Seminar der CAU zu Kiel, WS 2008/2009

[41] N. Makhijani, R. K. Sharma, J. B. Srivastava, Units in finite dihedral and quaternion group algebras, Journal of the Egyptian Mathematical Society, Volume 24, Issue 1, January 2016, pp. 5-7

[42] Neha Makhijani, R. K. Sharma, J. B. Srivastava, The unit group of finite group algebra of a generalized dihedral group, Asian-European Journal of Mathematics Vol. 7, No. 2 (2014) 1450034 (5 pages)

[43] G.E. Murphy, A new construction of Young's seminormal representation of the symmetric group, Journal of algebra 69, vol. 2, 1981, pp. 287-297

[44] Gordon James, Martin W. Liebeck, Representations and Characters of Groups, Cambridge University Press, 2001

[45] Macdonald, I. G., Symmetric functions and Hall polynomials. Second edition. Clarendon Press, Oxford, 1995

[46] I. B. S. Passi, D. S. Passman, S. K. Sehgal, Lie solvable group rings, Can. J. Math., Vol. XXV, No. 4, 1973, pp. 748-757

[47] D. S. Passman, Observations on group rings, Communications in Algebra, 5(11), 1977, pp. 1119-1162

[48] S. Perlis - G.L. Walker, Abelian group algebras of finite order, Trans. Amer. Math. Soc. 68, 1950, pp. 420-426

[49] R.S. Pierce, Associative Algebras, Springer-Verlag, New York, 1982

[50] Premet, A. A., Cartan subalgebras of Lie p-algebras, Izv. Akad. Nauk SSSR Ser. Mat. 50, 1986, number 4, pp. 788-800

[51] Robert C. Rhoades, Rank of symmetric matrices over finite fields, $math.stanford.edu/rhoades/FILES$

[52] Robinson, D.J.S., A course in the theory of groups, Springer-Verlag, New York, 1982

[53] Robinson, Geoffrey R., A bound on norms of generalized characters with applications, Journal of Algebra, 212, 1999, pp. 660-668

[54] Robinson, Geoffrey R., More bounds on norms of generalized characters with applications to p-local bounds and blocks, Bulletin of London Mathematical Society, 37, 4, 2005, pp. 555-565

[55] Robinson, Geoffrey R., On generalized characters of nilpotent groups, Journal of Algebra, 308, 2007, pp. 822-827

[56] Robinson, Geoffrey R., On the minimal norm of a non-regular generalized character of an arbitrary finite group, Bulletin LMS, 2010

[57] Jean-Pierre Serre, Linear Representations of Finite Groups (Graduate Texts in Mathematics), Springer, 1996

[58] Seligman, G.B., Modular Lie Algebras, Springer, Berlin, Heidelberg, New York, 1967

[59] S. Siciliano, Cartan subalgebras in Lie algebras of associative algebras, Communications in Algebra, Volume 34, Issue 12 December 2006 , pp. 4513 - 4522

[60] S. Siciliano, On the Cartan subalgebras of Lie algebras over small fields, J. Lie Theory 13, 2003, pp. 511-518

[61] M.C. Slattery, Computing character dgeress of p-groups, Journal of symbolic computation, vol. 2, 1986, pp. 51-58

[62] Benjamin Steinberg, http://math.stackexchange.com/questions/819466/the-division-algebras-arising-in-the-wedderburn-decomposition-of-a-finite-group

[63] Bernd Stellmacher, Hans Kurweil, Theorie der endlichen Gruppen, Springer-Verlag, 1998

[64] Stitzinger, E., Theorems on Cartan subalgebras like some on Carter subgroups, Trans. Amer. Math. Soc. 159, 1971, pp. 307-315

[65] H. Strade, R. Farnsteiner, Modular Lie algebras and their representations. Marcel Dekker, New York, 1988

[66] Charles J. Stuth, A Generalization of the Cartan-Brauer-Hua Theorem, Proceedings of the American Mathematical Society, Vol. 15, No. 2, April 1964, pp. 211-217

[67] Ya. Sysak, Associative rings and their adjoint groups, Arbeitsgruppe Gruppentheorie, Universität Mainz

[68] Tamburini, M.C. - Vdovin, E.P., Carter subgroups in finite groups, J. Algebra 255, 2002, pp. 148-163

[69] A. M. Vershik and A. Yu. Okounkov, A new approach to the representation theory of symmetric groups. II, arXiv.org ¿ math ¿ arXiv:math/0503040

[70] Vinroot, C. Ryan, Twisted Frobenius-Schur indicators of finite symplectic groups, Journal of Algebra, 293, 2005, pp. 279-311

[71] Vinroot, C. Ryan, A note on orthogonal similitude groups, Linear and Multilinear Algebra, 54(6), 2006, pp. 391-396

[72] Vinroot, C. Ryan, Character degree sums and real represenations of finite classical groups of odd characteristic, Journal of Algebra and Its Applications, 09, 633 (2010), pp. 633-658

[73] Dalla Volta, F. - Lucchini, A. - Tamburini, M.C., On the Conjugacy Problem for Carter Subgroups, Comm. Algebra 26, 1998, pp. 395-401

[74] S. Wirsing, Über separable Elemente in assoziativen Algebren, disserta diplomica, Hamburg, 2014

[75] S. Wirsing, Über Einheitengruppen modularer Gruppenalgebren, disserta diplomica, Hamburg, 2014

[76] S. Wirsing, Über die Struktur der Solomon-Tits-Algebren, disserta diplomica, Hamburg, 2014

[77] S. Wirsing, Maximal nilpotente Teilstrukturen I, disserta diplomica, Hamburg, 2015

[78] S. Wirsing, Maximal nilpotente Teilstrukturen II, disserta diplomica, Hamburg, 2015

[79] Wikipedia (englisch, deutsch), references for telephon-numbers and historical background of diverse mathematician

[80] R.L. Wilson, Cartan subalgebras of simple Lie algebras, Trans. Amer. Math. Soc 234, 1977, pp. 435-446

[81] D.J. Winter, On the toral structure of p-algebras, Acta Math. 123, 1969, pp. 70-81

[82] The GAP Group, GAP — Groups, Algorithms, and Programming, Version 4.2; Aachen, St Andrews, 1999, http://www-gap.dcs.st-and.ac.uk/ gap

[83] https://www2.bc.edu/ reederma/SL(2,q).pdf

[84] http://math.stackexchange.com/questions/908439/examples-of-division-algebras

[85] http://de.wikipedia.org/wiki/Lokaler Ring

[86] http://de.wikipedia.org/wiki/Symplektische Gruppe

[87] http://math.stackexchange.com/questions/38571

[88] https://people.kth.se/ boij/kandexjobbVT11/Material/pgroups.pdf

[89] http://math.stackexchange.com/questions/148817/representations-of-central-products

[90] http://math.stackexchange.com/questions/267708/ does-the-symmetric-difference-operator-define-a-group-on-the-powerset-of-a-set

[91] http://mathoverflow.net/questions/249661/cartan-subalgebras-of-matrix-algebras-over-fields-and-division-algebras

[92] http://mathoverflow.net/questions/251953/dimension-of-maximal-tori-in-division-algebras

Index

$(A \times A; \odot)$ zero extension of A, 15
$(K; L)$ field extension, 15
$(P(M); \cdot)$ power set of M with complex product $\cdot$ as operation, 13
$(P(N); \cap)$ power set of N with operation $\cap$, 13
$(P(N); \cup)$ power set of N with operation $\cup$, 13
$(P(N); \delta)$ power set of N with operation δ - symmetric difference, 13
$A(a,b)$ generalized quaternion algebra, 15
A/I factor algebra, 14
$A/rad(A)$ factor algebra by the nilradical, 16
A^K adjunction of a unit, 14
A° associated Lie algebra, 14, 34
$A^{n \times n}$ algebra of $n \times n$-matrices over A, 14
A^{op} or A^- inverse or opposite algebra of A, 15
A_n alternating group of degree n, 13
$Aug(KG)$ augmentation ideal of KG, 15
$B(n)$ Bell numbers, 77
$C(G)$ set of conjugacy classes of G, 37
$C_G(T)$ centralizer of T in G, 18
C_n or Z_n cyclic group of order n, 14
$D(n,K)$ diagonal matrices, 76
D_n Solomon algebra, 16
D_{2n} dihedral group of order $2n$, 13
$E(A)$ group of units of A, 14
$G = (A, \tau)$ generalized dihedral group, 167
$GF(p^n)$ finite field with p^n elements, 15
$GL(n,q)$ general linear group of degree n over $GF(q)$, 14
$GSP(2n,q)$ general similitudes group, 14
$J(A)$ Jacobson radical von A, 16
$K(a)$ smallest subfield in L containing a and K, 15
KG group algebra, 14
KM monoid algebra, 14
$K[t]$ polynomial algebra over K in the variable t, 15
$K\Pi_n$ Solomon-Tits algebra, 16
$N_G(T)$ normalizer of T in G, 18
$PSL(n,q)$ projective special general linear group of degree n over $GF(q)$, 14
$Q(A)$ quasiregular group of A, 14
Q_{4n} quaternion group of order $4n$, 13
$Rad(L)$ greatest solvable ideal, 47
$S(n,k)$ Stirling numbers, 42
SD_{2^n} semi-dihedral group of order 2^n, 13
$SL(n,q)$ special general linear group of degree n over $GF(q)$, 14
$SP(2n,q)$ symplectic group of degree $2n$ over $GF(q)$, 14
S_n symmetric group of degree n, 13
$U(n,q)$ unitary group of degree n over $GF(q)$, 14
$VSEP(A)$ fully separable elements of A, 75
$Z(A)$ center of an algebra, 18
$[a,b]$ commutator of a and b, 25

$\Omega_2(A)$ greatest elementary-2-abelian subgroup of A, 167
$\Phi(\cdot)$ Euler's totient function, 138
$\delta_{o,n}$ algebra of upper triangular matrices of $K^{n\times n}$, 16
$\delta_{u,n}$ algebra of lower triangular matrices of $K^{n\times n}$, 16
λ left-regular representation, 34
$\langle T\rangle_K$ K-linear span, 14
$\langle T\rangle_{A_1}$ unital subalgebra generated by T, 14
$\langle T\rangle_A$ subalgebra generated by T, 14
$\ltimes$ semidirect product of groups, 14
$\mathbb{C}$ complex number field, 15
$\mathbb{H}$ real quaternion algebra, 15
$\mathbb{N}$ natural numbers, 13
$\mathbb{N}_0$ natural numbers containing zero, 13
$\mathbb{Q}$ rational number field, 15
$\mathbb{R}$ real number field, 15
$\mathbb{Z}$ the set of integers, 15
$\mathcal{C}_{A°}$ the set of Cartan subalgebras of $A°$, 75
$\mathcal{T}_A$ the set of maximal tori of A, 75
ω_d primitive dth root of unity, 15
$\oplus$ direct sum of algebras, 14
$\otimes$ tensor product of algebras, 14
$\overline{T}$ sum of all elements of T in KG, 37
ρ right-regular representation, 34
$\times$ direct products, 14
$\wedge_n, \wedge$ product on Π_n, 42
$\wr$ regular wreath product of groups, 14
^{g} conjugation with g, 18
$ad(l)$ multiplication with l in a Lie algebra, 34
$cl(L)$ nilpotency class, 66
$core_G(U)$ the core of U in G, 38
eAe identical to $\{eae \mid a \in A\}$ for an idempotent e, 15
$gl(n, K)$ identical to $(K^{n\times n})°$, 15
id_M identity function on M, 71
$ind(D)$ index of D, 82

$k(G)$ quantity of all conjugacy classes of G, 37
$o(g)$ order of g in G, 17
$p(n)$ number of partitions of n, 42
$rad(A)$ nilradical A, 16
$s\delta_{o,n}$ algebra of strict upper triangular matrices of $K^{n\times n}$, 16
$s\delta_{u,n}$ algebra of strict lower triangular matrices of $K^{n\times n}$, 16
$V_0(\mathcal{L})$ Fitting null-component of V with resp. to $\mathcal{L}$, 149

abelian group
   main theorem, 17
algebra
   basic, 125
   construction of reduced, 137
   cyclic, 118
   derived, 215
   isomorphism of derived algebras, 218
   local, 147
   reduced, 125
   reversible, 132
   unitary, 18
Amitsur, 21
Artin, 49
associated Jordan algebra, 68
associated Lie ring
   nilradical, 45
   solvable radical, 45
associative algebra
   radical complement, 35
   solvable, 35

Brauer, 28

Cartan subalgebra, 66, 71
   associative subalgebra, 66
   central-simple algebra, 116
   conjugacy, 197
   dihedral group, 97
   dihedral groups, 200
   dihedral groups II, 201

direct products, 120
eAe, 106
Engel subalgebra, 149
factor algebra, 150
group algebra of hamiltonian group, 133
independence within same characteristic, 200
invariants, 198
invariants for constructions, 198
matrix algebra, 151
maximal tori, 75
number of, 79
quaternion algebra, 86
quaternion group, 100
quaternion groups, 202
reduced algebra, 127
self-centralizing, 102
semisimple algebra, 120
separable algebra, 120
separable factor algebra by the nilradical, 143
solvability, 68
solvable algebra, 91
solvable group algebras, 96
splitting algebras, 101
strategy, 144
strategy for dihedral groups, 144
strategy for solvable algebras, 144
substructure, 148
tensor product, 151
unique dimension based on maximal tori, 195
unique dimension for KP, 195
unique dimension for KS_3, 194
unique dimension for algebras with central nilradical, 193
unique dimension for associative algebras, 195
unique dimension for group algebras, 196
unique dimension for separable algebras, 192
unique dimension for separable group algebras, 192
unique dimension for solvable algebras, 192
weight, 197
zero-extension, 107

Cartan subalgebras
  associated Lie algebra, 74
  dihedral groups, 75
  division algebra, 82
  local algebra, 147
  quaternion groups, 76
  self-centralizing radical complements, 76
  Solomon algebra, 77
  Solomon-Tits algebra, 77
  splittable Lie algebra, 74
  triangular matrices, 77
Cartan-Brauer-Hua, 25

De Morgan, 141
Dedekind, 130
defect class, 42
derivation, 66
Division algebra
  finite abelian subgroups, 20
  finite subgroups, 21
  finite subgroups in positive characteristics, 20
  finite subgroups of uneven order, 21
  normal subgroup, 26
  solvable group of units, 27
  subnormal subgroups, 29

element
  fully separable, 34
  Engel, 149
field
  finite subgroup, 17
Frobenius-Schur indicator, 167
function
  centralizing, 75
  fully separable elements, 75

generalized Jordan decomposition
   adjoint representation, 35
   definition, 35
group
   Carter subgroups, 213
   derived subgroup, 37
   Fitting subgroup, 213
   Frobenius group, 170
   meta-cyclic, 21
   Z-group, 21
group algebra
   Dimension of maximal torus, 157
      first summation formula by Salvatore Siciliano, 157
   dimension of maximal torus
      $(ax+b)$-group, 170
   A_4, 166
   S_4, 166
   abelian groups, 160
   ambivalent group, 167
   central products, 181
   cyclic maximal subgroup, 179
   dihedral groups, 161
   direct products of two groups, 162
   extra-special group, 178
   extra-special p-groups, 161
   Frobenius groups, 170
   general similitude group, 169
   generalized dihedral group, 168
   groups of order p^3, 181
   groups of order p^4, 181
   large center, 179
   linear group, 168
   meta-cyclic group, 174
   minimal non-abelian p-groups, 165
   nilpotent groups, 163
   p-groups, 164
   pq-groups, 161
   quaternion groups, 162
   semi-dihedral groups, 162
   semidirect product with abelian normal subgroup, 172
   Slattery-algorithm, 164
   subgroups, 181
   symmetric group, 167
   symplectic group, 169
   upper and lower bounds, 158
   local, 147
   reduced and modular, 133
   reversible, 132
   semisimple and reduced, 132
   symmetric, 133

Hamilton, 129
Herstein, 20
Hua-identity, 25

induced character, 171

Jacobson, 122
Jordan decomposition, 34
   conjugacy class sums, 39

Lie algebra
   splittable, 73
Lie nilpotency
   group algebra, 112
Lie nilpotent
   central fully separable elements, 107
   central radical complement, 108
   division algebra, 81
   fully separable elements, 107
   group of units, 110
   solvable, 67
   tensor products, 109

Mackey, 172
Maschke, 37
maximal separable subfield, 81
maximal tori, 71
   associative subalgebra, 72
   dihedral group, 72
   division algebra, 82
   quaternion algebra, 86
   quaternion group, 72
   solvable algebra, 91

239

maximal torus
    central-simple algebra, 116
    direct products, 190
    factor algebra of nilpotent ideals, 175
    unique dimension, 191
Morita, 55

Nilradical
    adjunction of a unit, 55
    algebra with separable factor algebra by its nilradical, 51
    definition, 33
    direct products, 48
    factor algebra, 54
    group algebra, 56
    ideal, 54
    left ideal, 53
    matrix algebras, 54
    opposite algebra, 54
    right ideal, 53
    semisimple algebra, 48
    simple algebra, 47
    Solomon algebra, 42
    Solomon-Tits algebra, 42
    solvable group algebras, 37, 38
    solvable group algebras of dihedral groups, 39
    solvable group algebras of groups of order $2 \cdot p^n$, 38
    solvable group algebras of quaternion groups, 39
    subalgebra, 53
    tensor products, 56
    triangular matrices, 41
    zero-extension, 59
Noether, 19

outlook on series II, 213

Pierce component
    algebra, 101
    nilradical, 102
    radical complement, 102
polynomial

fully separable, 34
separable, 34
squarefree, 34

radical complement
    self-centralizing, 102
restricted Lie algebra, 71

Schur, 159
Schur-Zassenhaus, 37
Scott, 27
separable maximal subfield, 81
set partitions, 42
Shoda, Kenjiro, 183
Skolem, 19
solvable algebra
    nilradical, 36
solvable radical
    direct products, 48
    semisimple algebra, 48
    simple algebra, 47
strict maximal subfield, 117
Stuth, 28
subalgebra
    maximal commutative, 122
    maximal nilpotent, 128
    maximal solvable, 126
    unital, 18
sum of two squares theorem, 132
Sylow, 21

telephone numbers, 167
twisted Frobenius-Schur indicator, 168

Wallace, 37
Wedderburn, 18